submarine telecoms
FORUM

SPECIAL FEATURE:
SubOptic
www.suboptic.org 2016

Finance & Legal Edition

87

WEBSITE TRAFFIC: UNIQUE VISITS

52,205
9-15

50,999
10-15

50,713
11-15

DOWNLOADS TO DATE:

58,659

54,874

59,873

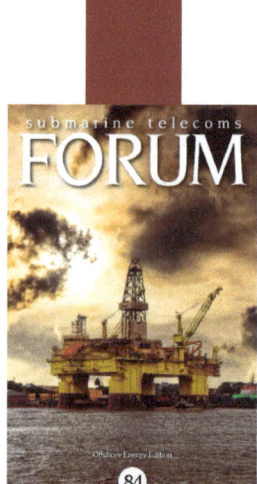

Issue #84 - Released 9-15

Issue #85 - Released 11-15

Issue #86 - Released 1-16

TOTAL HITS IN 2015:
10,303,895

55,109
1-16

55,980
2-16

47,612
12-15

520,542

522,927

350,894

Issue #15 - Released 8-15

Issue #16 - Released 11-15

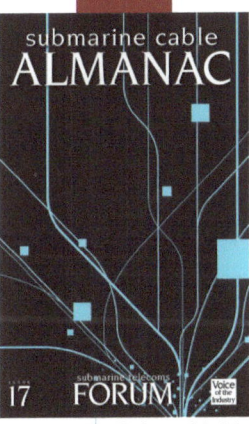

Issue #17 - Released 2-16

EXORDIUM

BY WAYNE NIELSEN

Welcome to Issue 87, our Finance and Legal, and Sub-Optic 2016 Special Edition!

I'm not one of the oldsters; my first SubOptic was only at Versailles in 1993 – with my family in tow.

In those days the exhibitors were relegated to a large tent outside the main proceedings. The technical papers were presented somewhere else, and because I pulled booth duty, I don't recall attending any of those sessions.

I do remember, however, attending the extraordinary evening fetes, and they were really something to appreciate; cocktails in La Louvre compliments of AT&T SSI, and a private baroque concert in Marie Antoinette's chapel thanks to ASN. My family and I had an amazing time in Versailles and most of the other SubOptics that followed, and at later conferences, I was even allowed to deliver a paper or poster or trophy.

Here we come to the next outstanding SubOptic venue, Dubai, which promises to be yet another important and memorable event. I'm not sure yet if I will be pulling booth duty, but I do know at a minimum I get to enjoy my progenies' first SubOptic poster presentation, which seems even more fitting years after my own first such session.

So, if you plan on attending SubOptic, please look me up; I'll either be the guy complaining about his tired feet, or the one behind the camera.

Wayne Nielsen is the Founder and Publisher of Submarine Telecoms Forum, and previously in 1991, founded and published "Soundings", a print magazine developed for then BT Marine. In 1998, he founded and published for SAIC the magazine, "Real Time", the industry's first electronic magazine. He has written a number of industry papers and articles over the years, and is the author of two published novels, Semblance of Balance (2002, 2014) and Snake Dancer's Song (2004).

+1.703.444.2527

wnielsen@subtelforum.com

IN THIS ISSUE...

ADVERTISER INDEX

News Now

➤ Algeria: 'Oran-Valence' Submarine Cable to Be Operational in February 2017

➤ Bandwidth Export to India Kicks Off

➤ Basslink Still Unsure Where Cable Fault is, Facing 'Phenomenal' Repair Costs

➤ Basslink Tasmanian Subsea Cable to be Repaired by May

➤ Brazil Reiterates Support to Angola Cables Initiative

➤ Cambodian Government Signs $70M Deal For Submarine Internet Cable

➤ DOCOMO PACIFIC Selects NEC to Build "ATISA" Submarine Cable

➤ Du Says May Take Longer to Repair Damaged Submarine Cable

➤ Gavin Tully Becomes Director of Submarine Solutions at Pioneer Consulting

News Now

- ➡ Internexa Extends Telefonica Submarine Cable Access

- ➡ Lightning Knocks Out Maya Submarine

- ➡ Linode: Major Cuts to Several Submarine Cables to Singapore

- ➡ New Hope for Hawaiki or Bluesky Cable After Pacific Island Leaders Meet in Auckland

- ➡ OFS Expands Ultra Long Haul Product Line with Introduction of TeraWave™ SCUBA Optical Fiber

- ➡ Omantel's Fibre-Optic Cable Set to Revolutionise Telecom in East Africa

- ➡ Repair Work Underway on TPG's PPC-1 Cable

- ➡ SEA-ME-WE 5 Completes Shore End in TI Sparkle's Landing Station in Catania, Sicily

- SEA-ME-WE 5 Submarine Cable Backers Announce Landings

- SEA-ME-WE 5 Undersea Cable Successfully Landed at Tuas

- SEA-ME-WE-5 Submarine Cable System Lands in Djibouti

- Slow Internet Connection Will Be Fixed by March 31, Says TM

- SLT Opens SEA-ME-WE 5 Submarine Cable Landing Station

- Submarine Cable Almanac Issue 17

- Submarine Cable Cut Lops Terabits Off Australia's Data Bridge

- Submarine Cable Work (Tasman Global Access) Begins at Raglan

- SubOptic 2016 – Register before March 31st 2016 to get the On-Time Rate.

News Now

➡ SubOptic 2016 – Venue Hotel Nearly Sold Out!

➡ SubOptic 2016: Exhibition Hall Has Sold Out – Sponsorship Packages Still Available

➡ Telefonica Creates New Infrastructure Company

➡ Telefonica to Lay New US-Brazil Undersea Cable

➡ The High-Speed SEA-ME-WE 5 Submarine Cable Lands at La Seyne-sur–Mer

➡ This Week in Submarine Telecoms February 1-5

➡ This Week in Submarine Telecoms February 29-March 1

➡ This Week in Submarine Telecoms February 8-12

➡ This Week in Submarine Telecoms March 14-18

➡ This Week in Submarine Telecoms March 7-11

Now offering video adverts for $500

FINANCE & LEGAL OUTLOOK

BY KIERAN CLARK

D espite global economic uncertainty, the submarine fiber industry has been enjoying something of a small resurgence in the last few months. Contract in force rates have gone up, money is changing hands, and more systems than ever are moving along and hitting development milestones. The next few years could be a welcome breath of fresh air across the industry.

Welcome to SubTel Forum's annual Finance and Legal issue. This month, we will take a look at the industry's current financial and legal status by presenting our most current data as tracked by the ever-evolving SubTel Forum database— where products like the Almanac, Cable Map, and the Industry Report find their roots. It has been a full year since our last look at the financial situation of planned systems around the world. New systems have been announced and planned systems have gone into service, while others have been delayed or changed. Quite a lot can happen in one year, and this year was no different.

Since our last Finance and Legal issue, we observed that of the planned systems announced to be ready for service in 2015, only 29 percent were actually accomplished. This is down a step further from 2014, which saw only 38

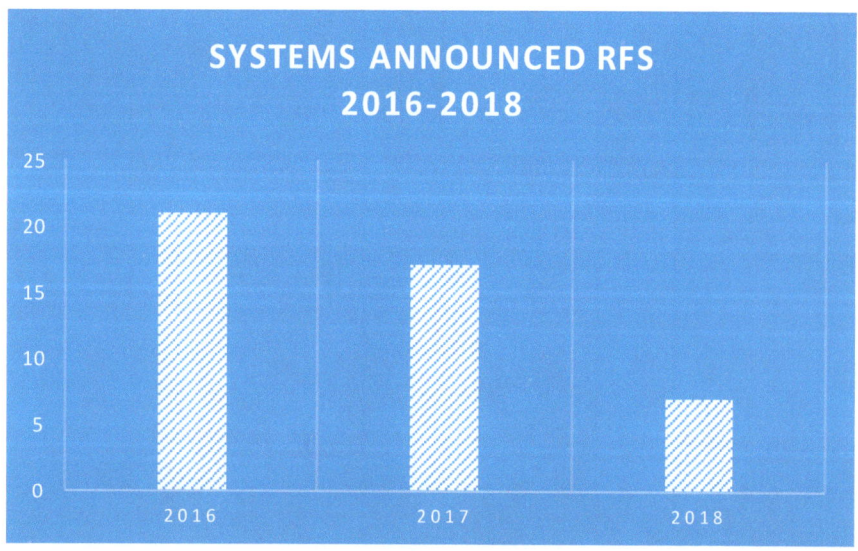

percent of systems move from planning to reality. Several of the remaining systems were either delayed to 2016 for various reasons, while 71 percent of the systems scheduled to be ready for service in 2015 simply died outright. Combined with only a net addition of four systems announced for 2016-2018 compared to a year ago, this seems like cause for concern. However, other factors in the cable development process seem to indicate otherwise.

Continuing to compare last year's numbers, the way systems are being financed maintains a shift towards single owners. Last year was the first time a noticeable trend leaning towards single owners was observed and the trend has continued this year. Over the next three years, only 33 percent of systems will be owned by a consortium, with the remaining 67 percent having a single owner. While consortium ownership reduces the financial risk to any single owner should a cable system fail, single ownership provides potentially greater flexibility and speed to the cable development process.

Scratching beneath the surface a bit, many of the new systems over the next several years are relatively smaller systems, serving very specific needs. A large portion of the systems planned for 2016-2018 connect various island

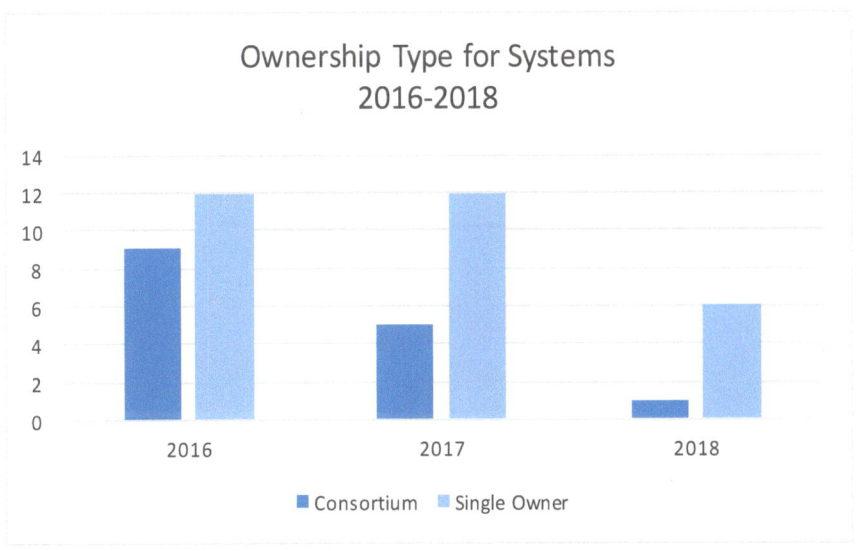

nations in the Pacific to existing international pipelines, or serve specific purposes such as connecting data centers within a cloud service provider's international infrastructure. With cloud service providers continuing to drive cable demand as a result of needing more control over the development process and a desire for faster system installations, this trend is expected to continue over the next several years.

While single owners continue to take over more of the submarine fiber industry's attention, the way a system is being financed has changed slightly from last year. Last March, we observed an even split between debt/equity and self-financed systems. This year, a new factor has been added throwing multilateral development banks into the mix. This shift continues to indicate that system owners and financiers are increasingly less willing to take on the uncertainty of debt/equity financing, and increasingly prefer having cash guaranteed up front. When the nature of many planned systems for the next few years is considered, these numbers make sense. So many of the planned systems serve very specific commercial and regional needs, weakening any business case attempting to justify financing based on selling future bandwidth.

Breaking the next three years down globally, the Austral-

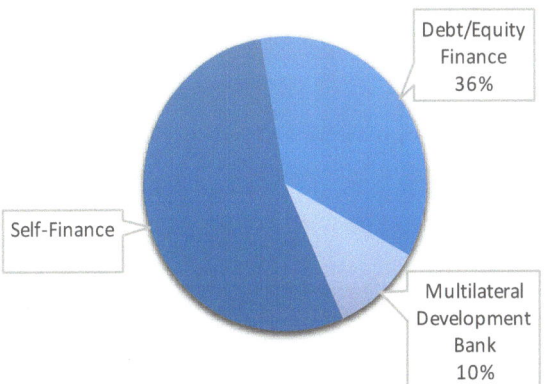

System Financing Type
2016-2018

Debt/Equity Finance 36%

Self-Finance

Multilateral Development Bank 10%

Asia region shows the most activity by far. Fully one-third of all planned systems for 2016-2018 will be developed in this region. This reflects the massive growth still occurring in Southeast Asia, as well as Australia's desire for increased route diversity. The Americas and EMEA regions account for 17 percent of future activity each, largely fueled by data center connectivity, and in part due to increased desire for routes to South America. Activity in the Transatlantic and Transpacific is almost entirely data center related, while the Indian Ocean Pan-East Asian region

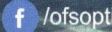 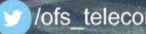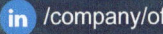

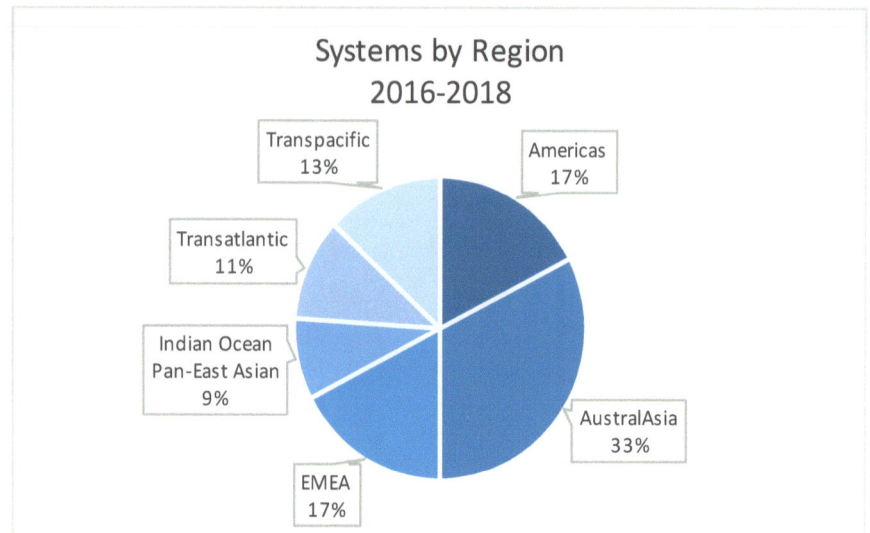

Systems by Region
2016-2018

- Transpacific 13%
- Americas 17%
- Transatlantic 11%
- Indian Ocean Pan-East Asian 9%
- AustralAsia 33%
- EMEA 17%

continues to benefit from being a crossroads between Asia and Europe.

With planned systems for the next three years up only slightly from a year ago, it may seem like the industry is continuing its dismal outlook of the last few years. However, when observing the contract-in-force (CIF) rate compared to last year, prospects begin to look much brighter. This is the true measure of a system's viability, and is usually a strong indicator that a system will be completed. At this time last year, only 17 per-

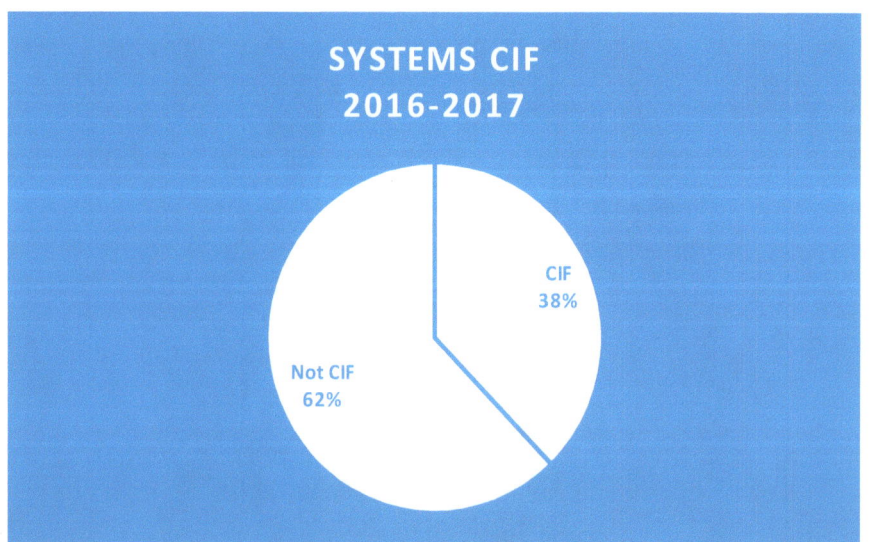

SYSTEMS CIF
2016-2017

- CIF 38%
- Not CIF 62%

cent of announced systems for 2015-2017 had reached this CIF milestone. This year, that number has more than doubled to 38 percent of planned systems for 2016-2018 hitting the CIF benchmark. Looking at 2016 alone, two-thirds of systems set to be ready for service are already CIF. With 2016 only three months in, these CIF rates are quite encouraging.

Getting money for a system continues to be the biggest hurdle in the cable development process. While this difficulty will almost certainly never go away, the outlook for the next few years is very promising. Much of the uncertainty that has been lingering over the industry has begun to subside, and money has started changing hands more optimistically than in recent times. While a system still needs a rock solid business case, banks and investors are more willing to put up capital to see a system come to fruition. With such a surge in confidence over the next few years, the submarine fiber industry promises to be very busy.

Kieran Clark is an Analyst for Submarine Telecoms Forum. He joined the company in 2013 as a Broadcast Technician to provide support for live event video streaming. In 2014, Kieran was promoted to Analyst and is currently responsible for the research and maintenance that supports the SubTel Forum International Submarine Cable Database; his analysis is featured in almost the entire array of SubTel Forum publications. He has 4+ years of live production experience and has worked alongside some of the premier organizations in video web streaming.

REGULATORY AND NATIONAL SECURITY CONCERNS FOR SUBMARINE CABLE CONSTRUCTION

BY ANDREW D. LIPMAN, ULISES R. PIN & CATHERINE KUERSTEN

Abstract:

International bandwidth usage has been growing at a compounded rate in recent years, forcing telecommunications companies to add bandwidth on numerous routes all over the world. This article examines some of the complexities of building international submarine cables including FCC and other licensing authorities.

The global demand for data is growing at an unprecedented rate. With more than 3 billion current Internet users, Internet traffic is anticipated to reach 14 gigabytes per capita by 2018. According to recent reports, international bandwidth usage has been growing at a compounded rate of over 82 percent over the past few years. Given their capacity, speed and security, submarine cables, carrying over 95 percent of international communications, remain the preferred medium for transporting data. As a result, telecommunications companies are scrambling to add bandwidth capacity on international and intercontinental submarine cable routes. Over the past four years, an estimated $11 billion has been poured into construction of new submarine cables, with approximately $1.8 billion in the Europe-Asia region alone.

Traditional telecommunications carriers are also no longer the only companies constructing and financing submarine cables. Internet and social media companies are building cable systems to move data and increase redundancies. Similarly, the financial trading industry is investing in subsea cables in order to increase speed and reduce latency. Cumulative submarine cable installations are currently projected to reach two million kilometers internationally by 2020.

Yet construction of a new submarine cable comes

with a multitude of challenges that can be barriers to market entry, including their exorbitant cost. For one, complicated and oftentimes opaque licensing and permitting requirements, or country-specific regulatory regimes, may delay or altogether impede deployment of new international subsea systems. Further, certain jurisdictions may have cybersecurity concerns regarding cable construction; these concerns remain even after completion of a cable and must be acknowledged

by cable operators in the interest of maintaining the integrity of the system and all communications transmitted over it.

International Deployment of Submarine Cables

Construction of a subsea cable system is a complicated process, subject to various licensing and permitting requirements that vary from country to country and impede development of new global systems. Many countries require at least two licenses: one for telecommunications and a separate for the submarine cable; but a myriad of other permits may be required as well, including environmental, construction and land use, and defense or national security authorizations. The rising rate of global terrorism makes issues such as physical security of and access to communications networks particularly salient. Countries that have suffered terrorist attacks, such as the United States, India, and the United Kingdom, tend to

have more stringent national security requirements for subsea cable builders. In all, the process for obtaining licenses can be both time-consuming and complicated.

These complications are only amplified in emerging markets, where cable systems are necessary for infrastructure support and critical to economic development. Countries in emerging markets now collectively account for 80 percent of the global population and 20 percent of the world's economic output; currently, Africa is the fastest-growing submarine cable market, with a compound annual growth rate of approximately 4.5 percent. Yet these jurisdictions tend to have less developed legal and regulatory regimes, leading to comparatively more opaque application and review processes. Some may subject applicants to bureaucratic impediments, and many enforce "localization" requirements for the use of local labor forces and other local professionals for construction or submission of licensing applications.

U.S. Regulatory Processes

Given its economic significance and its position as one of the world's largest content producers, a large number of new submarine cable systems land in the United States. Obtaining the requisite permits and licenses to land and operate a submarine cable in the United States can be an arduous process taking up to a year, if not more. Subsea cable builders must obtain licenses from the Federal Communications Commission ("FCC"); pass national a security review; and seek authorization from other federal, state, or local agencies depending on the nature and location of the cable. Furthermore, most permitting processes involve public consultation proceedings and may carry significant political implications. Sufficient lead time and knowledgeable and experienced advisors are critical for securing necessary approvals and timely commencement of operations.

FCC Approval

The FCC regulates submarine cable landings in the United States. Operators and constructors of submarine cables must obtain an FCC license pursuant to the Submarine Cable Landing License Act of 1921 and Section 1.767 of the FCC's Rules. The application process for a cable license may be completed in as little as 45 days from the date the application is put on public notice. However, in practice, the FCC licensing process tends to take significantly more than 45 days. Under FCC Rules, the application must be submitted by all entities that (1) own or control a U.S. landing station or (2) own or control a five percent or greater interest in the cable system and will use the U.S. segments of the cable system.

The application itself consists of: a description of the submarine cable, including the type and number of channels and its capacity; a specific description of the cable landing stations both in domestic and foreign territories; a map showing the geographic coordinates of all landing stations; a statement as to whether operations of the cable will be on a common or private carrier basis; and ownership information, including identification of all original owners of the cable, regardless of the amount of their ownership interests.

Additional licensing may be required if a licensee operates a submarine cable facility as a common carrier (*i.e.*, a telecommunications provider compelled the serve the public indifferently). In this case, an operator must obtain a service license to provide international telecommunications services under Section 214 of the Communications Act of 1934, as amended, commonly referred to as "Section 214 authorization." The Section 214 and cable license applications are submitted, and generally granted, at the same time.

Subsea cable owners are exempt from applying for a service license if they can show that they are a non-common carrier, which allows them to sell capacity on an individual case basis. Two conditions must be met for a sufficient showing of non-common carrier status: first, the owner must show that there are alternative common carrier facilities available on the cable route; and second, that there are no reasons implicit in the nature of the cable system operations requiring common carrier treatment. Non-common carriers do not need any FCC authority beyond the cable license for operation.

National Security Concerns

Because most subsea cable applications have some level of foreign ownership or participation, submarine cable landing license applications are also subject to review from "Team Telecom," an *ad hoc* task force comprised of the

Departments of Defense, Homeland Security, and Justice, including the Federal Bureau of Investigations. Though Team Telecom review is not statutorily mandated, it has plenty of precedent in the realm of submarine cable licenses. Further, Team Telecom frequently requests the FCC to defer granting a license application until completion of its review. As a result, the FCC will rarely grant a landing license in fewer than six months where Team Telecom review is required.

In general, the process is as follows: (1) the FCC provides copies of applications with foreign ownership to the Executive Branch for review; (2) Team Telecom asks the applicants a series of questions, commonly known as "triage questions," pertaining to issues such as equipment type, storage and security of call data and other physical security information, encryption key usage, and entities with access to the applicant's network and data; (3) the review generally culminates with the government

and applicants entering into a commitment letter or network security agreement addressing national security concerns. In order to streamline the process and expedite review, applicants may proactively contact Team Telecom before or upon filing at the FCC.

Network security agreements are critical to facilitating surveillance programs conducted by U.S. national security and law enforcement agencies, for example the National Security Agency, as well as for preventing foreign governments from gaining visibility into U.S. telecommunications and surveillance systems. Provisions of such agreements frequently include limitations on equipment types used, or requirements to establish a network operations center ("NOC") located domestically and operated by screened U.S. citizens.

Cybersecurity remains a growing concern both internationally and domes-

tically. In 2013, President Obama issued an executive order aimed at improving critical infrastructure cybersecurity, following which the National Institute of Standards and Technology, a sector of the Department of Commerce, released a Cybersecurity Framework promulgating a set of industry standards to reduce cyber risks to critical infrastructure. This framework, implemented by multiple industry sectors including the Department of the Treasury, was also integrated by the FCC through its Communications, Security, Reliability, and Interoperability Counsel (CSRIC). In March of 2015, CSRIC released a report recommending to the FCC actions for cybersecurity risk management, and to providers and operators guidance for operationalizing the framework.

Other Authorizations

Required federal authorization are not limited to FCC approval: submarine cable operators must also obtain a federal permit from the Army Corps of Engineers (the "Corps") pursuant to Section 404 of the Clean Water Act, which regulates discharges of dredged or fill materials into waters of the United States, and Section 10 of the Rivers and Harbors Act of 1899, which governs all work in or affecting navigable domestic waters. Where individual permits are required, the Corps must complete an environmental review under the National Environmental Policy Act.

In addition, authorizations from the state or municipality, independent of the Corps, may be required as well. These are specific to where the cable routes and lands. For example, some states require a coastal development permit, or approval from state lands commissions which exercise jurisdiction over tidelands and submerged lands adjacent to the coast and offshore islands of the states. States such as California have extremely complex and problematic permitting requirements including development of an Environmental Impact Report ("EIR"); projects that are controversial or heavily challenged may cause the EIR process to span years.

Looking Forward

Submarine cables remain an important point of focus for national security. For example, last fall, news sources reported U.S. intelligence and military concerns over Russian submarines and spy ships "aggressively operating" near undersea cables, fearing sabotage. Indeed, cybersecurity and data vulnerability are likely to be the primary considerations for U.S. and foreign national security authorities for the foreseeable future. While the Cybersecurity Framework has been well-received by agencies and industry alike, its implementation is still ongoing, and will likely continue to impact cybersecurity issues concerning subsea cable systems.

As cyberterrorism increases internationally, the process for obtaining national security approvals may only become more onerous. However, the U.S. submarine cable market remains open to foreign investment and activity. The government frequently allows for flexible and creative ownership and control structures in order to sufficiently address national security issues through a national security agreement or other security commitment.

Operators and constructors of submarine cables should not be dissuaded by the level of regulation internationally, particularly given the critical and essential nature of submarine cables. With the assistance of persons knowledgeable of specific regulatory regimes, many roadblocks may be circumvented. In the U.S., for example, the multi-layered process of cable authorization can be streamlined if applicants of requisite permits – generally the subsea cable sponsors or their suppliers, depending on the specific cable supply contract – convene a pre-application meeting with all necessary federal, state, and local agencies. Thus, it is crucial that parties plan well in advance and utilize advisors experienced in the legal and regulatory issues at stake in order to best prevent delays and expedite the process.

Mr. Lipman is the Chair in the Telecommunications Media and Technology practice group of Morgan, Lewis & Bockius LLP, a 2000-attorney international law firm.

Mr. Pin is a partner in the Telecommunications Media and Technology practice group of Morgan, Lewis & Bockius LLP, a 2000-attorney international law firm.

Catherine Kuersten's practice focuses on advising domestic and international technology companies on regulatory, corporate, and litigation matters within the sectors of telecommunications, privacy, and cybersecurity. Catherine also has experience in intellectual property litigation, in particular Section 337 litigation before the US International Trade Commission. She previously worked for the US Air Force Court of Criminal Appeals, the US International Trade Commission Office of the General Counsel, and National Public Radio.

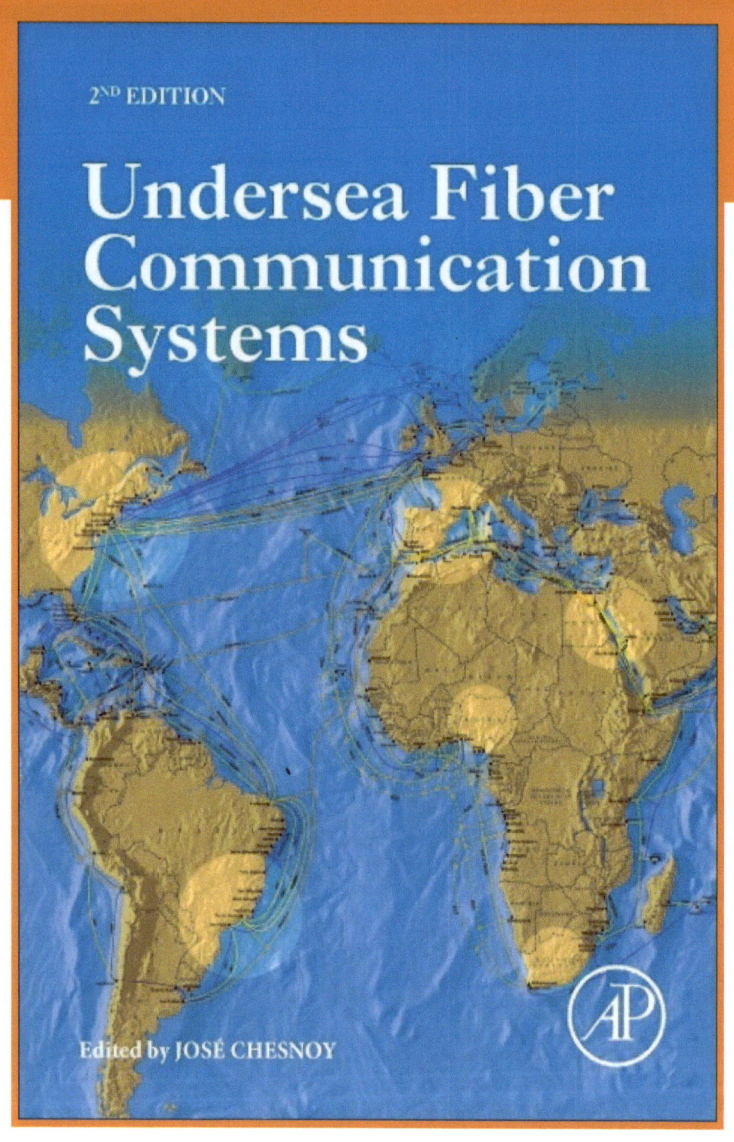

2ND EDITION

Undersea Fiber Communication Systems

Edited by JOSÉ CHESNOY

This comprehensive book provides both a high-level overview of submarine systems and the detailed specialist technical data for design, installation, repair, and all other aspects of this field.

Undersea Fiber

Communication Systems, 2e

Edited by: José Chesnoy
**Independent Submarine Telecom Expert,
former CTO of Alcatel-Lucent Submarine Networks**
With contributions of authors from key suppliers acting in the domain, such as Alcatel-Lucent, Ciena, NEC, TE-Subcom, Xtera, from consultant and operators such as Axiom, OSI, Orange, and from University and organization references such as TelecomParisTech, and Suboptic, treating the field in a broad, thorough and un-biased approach.

KEY FEATURES

- Features new content on:
    - Ultra-long haul submarine transmission technologies for telecommunications
    - Alternative submarine cable applications, such as scientific or oil and gas
- Addresses the development of high-speed networks for multiplying Internet and broadband services with:
    - Coherent optical technology for 100Gbit/s channels or above
    - Wet plant optical networking and configurability
- Provides a full overview of the evolution of the field conveys the strategic importance of large undersea projects with:
    - Technical and organizational life cycle of a submarine network
    - Upgrades of amplified submarine cables by coherent technology

submarine telecoms
FORUM

Sponsors Awards For
"Best Poster"
&
"Best Paper"
At

SubOptic
2016

www.suboptic.org

SUBOPTIC 2016 PREVIEW

SubOptic
2016
www.suboptic.org

Welcome from the HOST - Omar Jassim Bin Kalban, MD & CEO (E-marine)

On behalf of E-marine, I am honoured and delighted that the 9th edition of the SubOptic Conference, "SubOptic 2016" is being held in Dubai, United Arab Emirates from 18th to 21st April 2016. SubOptic will be 30 years old with the staging of the SubOptic 2016 in Dubai. This leading event in the submarine cable industry is being hosted for the first time in its 30 years history in the Middle-East Region (outside US & Europe) and E-marine is indeed proud to be the host in Dubai. It is the longest running event in the submarine cable industry with its various unique characteristics and organization.

Interestingly, the 30th SubOptic anniversary has coincided with the region's forming crossroads for links between the Asian and European continents and thus has become an essential hub for the undersea communication network in recent years.

Submarine cable systems carry most if not all international data through approximately hundreds of different cables linking all major continents and most countries with coastlines around the world. There are hundreds of thousands of kilometres of cable, deployed to a greater depths, all installed by companies and participants at the SubOptic conference.

The internet and so many aspects of global living from banking to education, health to entertainment, depend on being continuously and effectively on-line; on-line means connected to a submarine cable system. Most governments recognise their submarine cable systems as being vital to the national economy. Over the last two decades, UAE in particular and the Region in general has invested heavily in undersea cable infrastructure projects mainly in Telecom, Oil & Gas business sectors.

The theme for the SubOptic 2016 conference is "Emerging Subsea Networks – The World's Expanding Treasure" as the world becomes ever more connected and we become closer and closer globally. Given that the conference is held only every three years and the time cycle in our industry for major projects tends to be years rather than months, we would be able to identify and highlight the various challenges and future issues.

In summary, SubOptic exists because it is the one conference event that is run by the industry for the industry.

This time the conference will take place at the Conrad Hotel, Dubai, UAE. Dubai is strategically located on the Eastern coast of the Arabian Peninsula, in the south west corner of the Arabian Gulf.

With year-round sunshine, intriguing deserts, beautiful beaches, luxurious hotels and shopping malls, fascinating heritage attractions and a thriving business community, UAE & Dubai receives millions of leisure and business visitors each year from around the globe.

As the host of SubOptic 2016 event, E-marine looks forward to seeing you all in Dubai, UAE.

Welcome from the President of the SubOptic Executive Committee

The SubOptic conference and convention comes around only once every three years and it is the premier event for the submarine cable industry – an event put on by the industry for the industry. On this occasion on behalf of the SubOptic Executive Committee I welcome you to Dubai for the ninth event in our series, SubOptic 2016, hosted by E-marine PJSC. This will be the first time that a SubOptic event has been held in the Middle East and it is quite apt that on our 30th anniversary we have ventured into this region, which is a crossroads for different cultures and a critical hub in the global undersea telecommunications network.

SubOptic is the most important conference for the industry and I am sure it will continue to enjoy the success of past years. It is the one event that encourages all people from the industry and community to get together and discuss issues that are critical to our business.

As in previous events we have shaped the programme to provide maximum value to the industry and to encourage attendance to the varied and broad range of presentations. These have been organised by our Programme and Paper Committees, led respectively by Alice Shelton from Alcatel Submarine Networks and Mark Andre from Orange, whose structure has followed the pattern initiated at SubOptic 2013.

These committees are formed entirely by volunteers, including nearly 60 abstract reviewers - and it is only by their hard work and dedication that we can maintain the professional and independent nature of the event and make it truly - "For the Industry, by the Industry".

With submarine system based networks being ever more critical to the global economy I look forward to your full participation in an event which will explore the "Emerging Subsea Networks" and really demonstrate that they are "The World's Expanding Treasure".

I look forward to meeting you whilst you are here.

Yves Ruggeri

Introduction from the Chairman of the Programme Committee

It is getting close to SubOptic 2016 and I'm pleased to say the conference programme is close to being finalized. When writing the previous introduction to the Call for Papers it looked a daunting task but thankfully with the help of the Programme Committee and nominations for topics and presenters from the industry, the programme has come together relatively smoothly.

Before deciding on the programme structure for SubOptic 2016, a review of all the feedback received after SubOptic 2013, either via the survey or via emails or comments, was done. This review looked at improvements that could be done either in the structure of the programme or in the content expected. The survey, completed by more than 180 people, provided a lot of useful information. There were more than 100 replies on plenary sessions and approximately 50 replies on each of the other sessions, including each of the 16 oral sessions.

The ratings on the conference overall were very positive; >70% responded excellent/very good; >90% responded excellent/very good/good.

But looking at the comments, it was clear that there were many people who felt there was too much in parallel and that we should try to avoid three parallel oral sessions, cutting back the total quantity of oral presentations and being more selective in choice of these presentations. It was also felt that the poster session was too crowded in space and that there were too many posters to possibly get round in one session.

To answer these points the programme for SubOptic 2016 is based on two parallel oral sessions. We will have 14 sessions in total, so a maximum of 70 oral presentations. For the posters, we did initially plan to have two poster sessions. However the quantity and quality of abstracts received was such that we were able to fit all requests within the one poster session.

We plan also to celebrate 30 years of SubOptic. When the first SubOptic conference was in held in Versailles in 1986, optical submarine cables were in their infancy. The World Wide Web was perhaps somewhere on a drawing board and the driver for submarine systems was purely voice traffic. All systems were regenerator based with a single fibre pair having a capacity of 280Mbit/s.

We have since introduced optically amplified systems, originally based on a single wavelength per fibre pair, now deployed as dense Wavelength Division Multiplexed (DWDM) systems with in excess of 100 wavelengths per fibre, with individual channel bit rate capacities of 100 Gbit/s and more, an increase in capacity of more than1,000 times each year.

In 1986 systems were procured and installed by a small group of companies, all based essentially around monopoly players, many of them government owned, from vendors who were considered the national champions in their country. It wasn't until 1998 at our third event that the first private company initiatives were being discussed as deregulation allowed the introduction of systems based upon this approach. This led the industry into the dot.com mode with private companies installing systems as if the fantasies of the internet based companies were actually real, or using the same traffic requirements to underpin the business case for parallel systems. Somehow we survived the burst of the dot.com bubble.

The nature of the traffic has also changed, with voice being a very small portion of the traffic today. Internet based traffic now predominates and this has led new players such as Google, Facebook and Microsoft to enter the market as cable owners to service their own needs. This change in traffic influences the way that traffic flows as it is no longer balanced and is mostly dependent upon where the majority of content is generated.

So over the past 30 years, the time that SubOptic has existed, the industry has seen many fundamental changes and SubOptic has been an observer and a recorder of these changes. We will be looking back over this time with a session dedicated to the 30th anniversary of SubOptic.

I look forward to seeing you in Dubai in April.

Alice Shelton

Introduction from the Chairman of the PAPERS Committee

I feel honoured to have been appointed as SubOptic 2016 Papers Chairman, supporting Alice who leads the Programme Committee and had my role at the previous SubOptic edition. So far so good! We have achieved some critical milestones in the past two months and we are now heading at finalizing the last details of the oral sessions and posters. The "Call for Papers" for SubOptic 2016 was launched in May 2015 and the deadline for submission was extended only once! The Review Committee worked on the abstracts for two months, scrutinizing the descriptions of the papers, making sure they were provided under the right topic area, requesting clarifications to authors, etc. A long process and telling figures: 59 reviewers, 3722 individual scores. A fair selection process too: all reviewers ranked their papers with no indication about the author or the company. In December 2015, the six Vice Chairmen made their recommendations for SubOptic 2016 and notifications were sent to all authors before the Christmas break. I guess that some well-deserved free time was used for the conference...as the deadline for submitting the final version of the paper was 31st of January 2016. Extended only once also! Despite the unexpected number of authors withdrawing their paper, most of the 11 oral sessions have now their papers labelled and their chairpersons identified; a new family of good-will people! The preparation of the poster event is also 99% complete (never say 100% to remain on the safe side).

I would like to thank very much Mohamed, Graham, Peter, Edwin, Stephen and Guillaume for accepting my proposition to lead a review team and bring their experience/knowledge for the benefit of the event. Thank you also for the ongoing support. There will be a drink at Dubai.

All my gratitude to the reviewers too, it's never an easy task to score multi-pages paper based on a few lines. There can't be a drink at Dubai, I'm sure you can understand, but I truly thank you and appreciate your efforts.

Last but not least, a warm welcome to the oral sessions chairpersons for this SubOptic 2016 edition. I'm confident that they will bring their enthusiasm and make these sessions a great place for learning, asking, thinking, challenging the audience and the authors themselves!

Mark Andre

MASTERCLASS / TUTORIALS

1. Extending Capacity and Reach - Demand and Supply

We are pleased to have confirmed a masterclass tutorial session to be provided by Professor Polina Bayvel and researchers from the Optical Networks Group at University College London (UCL), currently engaged in the large EPRSC funded UNLOC project.

UNLOC - UNLocking the capacity of Optical Communications - is a programme funded by the UK Engineering and Physical Sciences Research Council (EPSRC) to research the future of optical communication systems.

It is recognized that global communication systems are rapidly approaching the fundamental information capacity of current optical fibre transmission technologies. To date, growth has been supported by falling cost/bit but by 2020 we may be running out of capacity. Capacity will become a rare commodity.

This will have a devastating impact on the economy, causing bandwidth to be rationed or prices to increase: either would be devastating for economic growth.

To set the scene for the presentation of the new approaches to unlock future capacity beyond the limits of current technology and tailored to the nonlinear optical channel, the first part of the session will describe the needed future scaling of submarine networks. Vijay Vusirikala and Valey Kamalov from Google will present on the drivers and trends for traffic growth for large data centre and cloud operators. This first part will also discuss how the business models, traffic growth pattern and diversity requirements are different from traditional subsea network models. They will also outline the value of open subsea cables and continued innovation in wet plant design and deployment.

Dr Robert Maher, Dr Dom Lavery and Professor Polina Bayvel, from UCL's Optical

Time	Monday 18th April	
Exhibition	**6:00 pm - 8:00 pm**	
8:00 - 9:00 am		
	Conference Room C	Conference Room D
9:00 - 9:30 am	Registration 8:00 am - 6:00 pm	
9:30 - 10:00 am		
10:00 - 10:30 am		
10:30 - 11:00 am		
11:00 - 11:30 am		
11:30 - 12:00 pm		
12:00 - 12:30 pm	Lunch	
12:30 - 1:00 pm		
1:00 - 1:30 pm	**Masterclass 1 EXTENDING CAPACITY & REACH**	**Masterclass 2 LEGAL INNOVATION & JURISDICTION CREEP**
1:30 - 2:00 pm		
2:00 - 2:30 pm		
2:30 - 3:00 pm	**Masterclass 3 POWER BUDGETS**	**Masterclass 4 COMPETING SEABED USERS**
3:00 - 3:30 pm		
3:30 - 4:00 pm		
4:00 - 4:30 pm	Break	
4:30 - 5:00 pm	**Masterclass 5 MYTHBUSTING**	**Masterclass 6 OIL & GAS - PROJECTS & TECHNOLOGY**
5:00 - 5:30 pm		
5:30 - 6:00 pm		
6:00 - 6:30 pm	**Welcome reception**	
6:30 - 7:00 pm		
7:00 - 7:30 pm		
7:30 - 8:00 pm		

Time	Tuesday 19th April	
Exhibition	9:00 am - 5:30 pm	
8:00 - 9:00 am	Registration 8:00 - 9:00 am	
	Conference Room C	Conference Room D
9:00 - 9:30 am	Opening ceremony (Plenary C & D)	
9:30 - 10:00 am	Keynote 1 - DR BASTAKI (C & D)	
10:00 - 10:30 am		
10:30 - 11:00 am	Break	
11:00 - 11:30 am	Paper session TU1A THE FIBRE CORNER	Paper session TU1B CABLE FINANCING AND CONSORTIUM ENVIRONMENT
11:30 - 12:00 pm		
12:00 - 12:30 pm		
12:30 - 1:00 pm	Lunch	
1:00 - 1:30 pm		
1:30 - 2:00 pm	Paper session TU2A UNREPEATERED APPLICATIONS	Round table 1 PROJECT FINANCE
2:00 - 2:30 pm		
2:30 - 3:00 pm		
3:00 - 3:30 pm	Break	
3:30 - 4:00 pm	Paper session TU3A UNDERSEA TECHNOLOGIES	Paper session TU3B THE MARINE CHAIN
4:00 - 4:30 pm		
4:30 - 5:00 pm		
5:00 - 5:30 pm	Workshop 1 - Open networks - the practical reality	
5:30 - 6:00 pm		
6:00 - 6:30 pm	End of day 6:00 pm	
6:30 - 7:00 pm		
7:00 - 7:30 pm		
7:30 - 8:00 pm		

Time	Wednesday 20th April	
Exhibition	**9:00 am - 5:30 pm**	
8:00 - 9:00 am		
	Conference Room C	Conference Room D
9:00 - 9:30 am	**Keynote 2 - Gerd Leonhard (Plenary C & D)**	
9:30 - 10:00 am		
10:00 - 10:30 am	**Round table 2 - The case for cables (C & D)**	
10:30 - 11:00 am		
11:00 - 11:30 am	Break	
11:30 - 12:00 pm	**Paper session WE1A MANAGING SYSTEM LIFETIME**	**Paper session WE1B MARINE ACTIVITIES**
12:00 - 12:30 pm		
12:30 - 1:00 pm		
1:00 - 1:30 pm	Lunch	
1:30 - 2:00 pm		
2:00 - 2:30 pm	**Paper session WE2A NETWORK TOPOLOGY**	**Paper session WE2B MARINE ASSETS AND OBSERVATORIES**
2:30 - 3:00 pm		
3:00 - 3:30 pm		
3:30 - 4:00 pm	**Poster session**	
4:00 - 4:30 pm		
4:30 - 5:00 pm		
5:00 - 5:30 pm		
5:30 - 6:00 pm		
6:00 - 6:30 pm	**End of day 6:00 pm**	
6:30 - 7:00 pm		
7:00 - 7:30 pm		
7:30 - 8:00 pm		

Time	Thursday 21st April	
Exhibition	9:00 am - 5:30 pm	
8:00 - 9:00 am		
	Conference Room C	Conference Room D
9:00 - 9:30 am	Workshop 2 - Extending system life (Plenary C & D)	
9:30 - 10:00 am		
10:00 - 10:30 am	Round table 3 - East meets West, a regional leader forum (C & D)	
10:30 - 11:00 am		
11:00 - 11:30 am	Break	
11:30 - 12:00 pm	Paper session TH1A LINE DESIGN	Paper session TH1B PROJECTS AND IMPACTS
12:00 - 12:30 pm		
12:30 - 1:00 pm		
1:00 - 1:30 pm	Lunch	
1:30 - 2:00 pm		
2:00 - 2:30 pm	Paper session TH2A WET AND DRY TECHNOLOGIES	Paper session TH2B THE LAST SESSION
2:30 - 3:00 pm		
3:00 - 3:30 pm		
3:30 - 4:00 pm	Break	
4:00 - 4:30 pm	Keynote 3 - 30th anniversary of SubOptic (C & D)	
4:30 - 5:00 pm		
5:00 - 5:30 pm	Closing ceremony	
5:30 - 6:00 pm	Free	
6:00 - 6:30 pm		
6:30 - 7:00 pm	Gala dinner	
7:00 - 7:30 pm		
7:30 - 8:00 pm		

Networks Group, will then discuss a range of promising digital signal processing techniques for unlocking the capacity of optical communications.

The success of next-generation multi-terabit/s optical transmission networks hinges on the development of optical transceiver subsystems that provide a significantly lower cost-per-bit than the extremely successful 100 G predecessors they will be replacing. In addition, this must be achieved within the unique power constraints inherent to submarine optical networks, while simultaneously maintaining or indeed increasing transmission reach. To achieve this objective, significant research focus has been placed on the generation and reception of higher order modulation formats, such as dual polarisation 16QAM and 64QAM, which have already been demonstrated in laboratory based trans-oceanic transmission systems up to 10,000km.

However, as the order of the modulation format increases, a corresponding increase in launch power is required to achieve an acceptable signal-to-noise ratio at the end of the transmission link. Higher powers inevitably result in significant signal distortion arising from fibre nonlinearity, which, if uncompensated, will limit the throughput of future optical transport systems. The masterclass tutorial will review different nonlinearity mitigation techniques that will enable the transmission of higher order modulation formats over submarine systems which can significantly increase both reach, capacity and spectral efficiency in submarine optical fibre systems.

Topics that will be covered:

- Scaling Submarine Networks: Datacentre Operator Perspective

- Overcoming and using nonlinearities to maximise system capacity

- Design of transceivers with optimal modulation, coding and DSP

Presenters:

- Valey Kamalov and Vijay Vusirikala (Google)

- Dr Robert Maher, Dr Dom Lavery and Professor Polina Bayvel (University College London)

Moderator: Steve Grubb (Facebook)

2. Legal Innovation & Jurisdiction Creep

Two separate topics will be covered during this 90 minute masterclass session.

A - Legal Innovation

Covering the legal implications of the new market conditions such as sales of spectrum not capacity and updates on many other legal matters for private and consortium cables.

Topics that will be covered:

- Legal implications of new market conditions (sales of spectrum not capacity)

- Sale and leaseback of submarine networks and capacity

- Resale restrictions in capacity contracts etc

- Protecting a purchaser of IRUs on a consortium cable from the insolvency of the seller

- Turning a private cable into a consortium

- Creating a new sales channel for a consortium

Presenter:

- Mike Conradi (DLA Piper)

B – Jurisdiction Creep

A masterclass on marine jurisdictional problems for submarine cables starting with an explanation of the rights of, and claims by, coastal states to the different maritime zones, describing the many problem areas where disputes can occur that can significantly delay and put at risk a submarine cable project, concluding with a practical section focused on remedies.

REACH into the future

Hexatronic is a strong partner for submarine cable solutions worldwide. Our highly skilled team has over 25 years of experience in submarine cable technology.

Meet us at SubOptic in Dubai, 18 - 21 April, Booth 36.

Topics that will be covered:

- Rights of, and claims by, coastal states to maritime zones

- Territorial sea, contiguous zone, EEZ, continental shelf, high seas, and the Area

- Authorities control beyond the off-shore 12 mile territorial waters – where is it happening?

- Varying requirements for securing landing licences and operational permits in different regions of the world

- What impact does this have to submarine cable projects on permits, timescales, costs etc

- Problem areas covering:

  - Permissible but harmful assertions of jurisdiction

  - Excessive assertions of jurisdiction inconsistent with international law

  - Failure to coordinate marine activities and protect submarine cables

SubOptic 2016
www.suboptic.org

- Failure to address emerging technologies and industries

- Failure to plan adequately for future submarine cable development

- Lack of private right of action under international law

- Lack of UNCLOS framework for resolving territorial disputes

- Practical solutions

Presenter:

- Kent Bressie (HWG)

Moderator: Carl Osborne (TATA Communications)

3. Optical Power Budgets – the key to the supply or upgrade of any submarine system?

In April 2015 Tony Frisch from Xtera invited anybody interested to join a Working Group on Optical Power Budgets. Two months later the Working Group had 12 members, all people with a wealth of experience in the industry, representing turnkey and upgrade suppliers, system owners, traditional carriers and

OTT players. A demonstration of the willingness of the industry to collaborate on a sensitive topic, the Working Group has been running since June and is ready to present its findings at SubOptic 2016.

Optical power budgets are key to the supply or upgrade of any submarine system, yet they sometimes appear – or are made to appear? – complex and difficult to understand. In reality, the essential principles are not that difficult to understand, even if deriving some of the numbers requires sophisticated tools and experience. The objective of the tutorial is de-mystify the essential processes, explain what the numbers mean and to show when it is necessary to go beyond approximations that are often used.

The masterclass will explain as simply as possible the essentials of how a power budget works and will clarify the essential similarities and differences between the two ITU templates which are used to present them. After this, a number of important refinements which aim to improve the accuracy of the budget will be explained. Finally there will be some discussion of how to compare or check budgets including real measurable parameters and how one might create and validate a budget for an "open system," where the Line is supplied independently of Terminal equipment.

Participating in the Working Group were experts from AJC, Alcatel-Lucent, AT&T, Ciena, Orange, Microsoft, Tata, Telstra, Verizon and Vodafone.

Topics that will be covered:

- Introduction – use in De-sign, Bidding and Acceptance

- Fundamentals – with a few simplifications

- How it applies to New builds, Upgrades, Unre-peatered Systems

- How to construct a Power Budget ITU-T formats Old and New ("for coherent systems")

- How it feeds into Acceptance

- Refinements – removing the simplifications – handling ROPA or DRA

- Discussion – how to do comparisons / sanity checks – how to handle "open systems"

Presenters:

- Priyanth Mehta (Ciena)

- Jamie Gaudette (Microsoft)

Moderator: Tony Frisch (Xtera)

4. Competition for Seabed Use - The Impact on the Planning, Installing and Maintenance of Submarine Cables

The primary objective of the masterclass is to address the increasing competition for seabed use in an increasingly complex regulatory environment and the impact on submarine telecommunications cables from a planning, survey, installation and maintenance perspective.

Masterclass topics will include:

- Emerging impacts from renewable energy

- Growing impacts from subsea mining

- Increasing demands from the hydrocarbons industry

- Coastal developments

- Marine parks and Marine Protected Areas

- Increase in protected marine habitats and designated sites of habitat concern

- Future regulation in Areas Beyond National Jurisdiction

Presenters:

- Kent Bressie (HWG)

- Tony Fisk (Pelagian)

- Kate Panayotou (GHD Sydney)

- Kai Schmidt (DTAG)

Moderator: Graham Evans (EGS)

5. MythBusters 2: Revenge of the Cable Myths

Topics that will be covered:

- The mythbusters are back with a brand new set of subsea cable myths to explode! After surviving the 2013 Paris SubOptic event with all extremities intact, the team of intrepid pseudo-scientists return to some of the most persistent myths in our industry:

- Internet traffic is doubling every two years.

- Bandwidth prices will eventually have to go back up.

- And the ever-popular: a majority of Internet traffic is "adult content".

- In addition to revisiting these myths and testing new ones, the mythbusters will also cast their gaze back further in time, to see which of the industry truisms that guided cable builds 30 years ago are busted … and which hold true.

Presenters:

- Alan Mauldin and Tim Stronge, with special guest star, David Ross

Moderator: David Ross

6. Oil & Gas – Projects & Technologies

Again a two part masterclass covering commercial and technical aspects of Oil and Gas projects and technologies.

The fibre optic concept for offshore platforms is no longer a new concept, but the industry has been knocked back recently by the significant drop in oil prices. The first part of this masterclass will discuss how we can approach and sometimes re-structure the current opportunities to accommodate a low oil price without impacting the key criteria of safety and reliability.

Topics that will be covered:

- The fibre optic concept for offshore platform has passed the tipping point and it is moving into the main stream for a number of reasons.

- The industry is facing some headwinds in the form of lower oil prices and high costs.

- How can we approach and sometimes re-structure the current opportunities to accommodate a low oil price ?

- Low cost does not mean low quality or lower safety standard

- What are the Cost, risk, threat not to allow these projects to happen ?

This commercial presentation will be complemented by a presentation on technologies covering Cable System Requirements and Technical Solutions for FO Subsea cables for the Oil & Gas Industry

Topics that will be covered:

- Introduction
  - o FO in offshore applications
  - o Comparison with alternative solutions
- Specifications and Requirements
  - o Standards
  - o Expectations and requirements specific to the Oil&Gas industry
- Typical Delivery Scope
  - o FO cables
  - o Accessories, in particular sub-sea terminations, riser solutions and top-side hang-off termination
  - o Installation reels
- Challenges
  - o Ease of installation
  - o Reliability
  - o Wet mate connectors (optical, electrical signal and electrical power)
- Example projects

Presenters:

- Pierre Tremblay (OSI)

- Inge Vinermyr (Nexans)

Moderator: Johann Benard (ASN)

ORAL/POSTER SESSIONS

Oral Presentation Sessions

Tuesday 19th April – Oral Sessions TU1A & TU1B 11.00AM – 12.30pm

ORAL SESSION TU1A – THE FIBRE CORNER

Session Chair – Stuart Barnes (Xtera)

TU1A – 1 Reduction in splice loss between fibers with dissimilar effective areas

Sergejs Makovejs (Corning)

TU1A – 2 Ultra-Large Effective Area Fibre Performances in High Fibre Count Cables and Joints. A new Technical Challenge

Florence Palacios (Alcatel-Lucent Submarine Networks)

TU1A – 3 Ultra-low loss Pure-silica-core fiber for capacity expansion

Takemi Hasegawa (Sumitomo Electric Industries)

TU1A – 4 Optical Fibre Fatigue And Submarine Networks Reliability: Why So Good?

David Walters (Independent Consultant)

TU1A – 5 Cable and Splice Performance of 153um² Ultra Large Area Fiber for Coherent Submarine Links

Ole Levring (OFS)

ORAL SESSION TU1B – CABLES FINANCING AND CONSORTIUM ENVIRONMENT

Session Chair – tbd

TU1B – 1 Financing Opportunities and Challenges Facing Submarine Cable Projects

Andrew Lipman (Morgan Lewis)

TU1B – 2 Fixing the Financing Shortfall for Next Generation Submarine Cable System Owners

Takeshi Takiguchi (NEC Corporation)

TU1B – 3 What Future For the

Consortium Model?

Joel Saltsman (Orange)

TU1B – 4 Cost Sharing Model for Future System Upgrade

Rayan Alsaedi (Saudi Telecom Company)

Tuesday 19ᵗʰ April – Oral Sessions TU2A 1.30pm – 3.00pm

ORAL SESSION TU2A – UNREPEATEReD APPLICA-TIONS

Session Chair – Reja Mateen (British Telecom)

TU2A – 1 Evolution of Repeaterless Systems Architectures

Philippe Perrier (Xtera)

TU2A – 2 Ultra-Low Loss Fiber and Advanced Raman Amplification Deliver Record Unrepeatered 100G Transmission

Ian Davis (Corning)

TU2A – 3 More than 30dB Budget Improvement in Unrepeatered 100GHz Links

Serguei Papernyi (MPB Communications Inc.)

TU2A – 4 Amplification technologies supporting upcoming modulation formats in unrepeatered links

Lutz Rapp (Coriant)

TU2A – 5 Enabling fibre and amplifier technologies for submarine transmission systems

Benyuan Zhu (OFS)

Tuesday 19ᵗʰ April – Oral Sessions TU3A & TU3B 3.30AM – 5.00pm

ORAL SESSION TU3A – UNDERSEA TECHNOLO-GIES

Session Chair – Olivier Courtois (Alcatel-Lucent Submarine Networks)

TU3A – 1 3D Printing Submerged Equipment

Adrian Jarvis (Huawei Marine Networks)

TU3A – 2 Extreme and Fatigue Analyses of a Dynamic Fiber Optic Riser

Muhammed Tedy Asyikin (Nexans)

TU3A – 3 Highly efficient submarine C+L EDFA with serial architecture

Douglas Aguiar (Padtec)

TU3A – 4 Technology for C+L Undersea Systems

Stuart Abbott (TESubcom)

TU3A – 5 Application and Benefits of Raman-Enhanced Amplification Schemes in Tomorrow's Optical Submarine Systems

Markus Nölle (FhG-HHI)

ORAL SESSION TU3B – The Marine Chain

Session Chair – Matteo Gumier (Alcatel-Lucent Submarine Networks)

TU3B – 1 Bigger Isn't Always Better: Focusing on Pertinent Desktop Study Content

Nancy Poirier (IT International Telecom)

TU3B – 2 Route Clearance for Hibernia Express and the findings

Alasdair Wilkie (Hibernia Networks)

TU3B – 3 Regulatory Challenges of Project Implementation – India Case Study

Nick Smith (Alcatel-Lucent Submarine Networks)

TU3B – 4 Global Trends in Submarine Cable System Faults

ME Kordahi (TESubcom)

TU3B – 5 The Benefits of Recycling the Right Way

Dec Wallace (BT), Arnaud Louw (Mertech Marine)

Wednesday 20th April – Oral Sessions WE1A & WE1B 11.30AM – 1.00pm

ORAL SESSION WE1A – MANAGING SYSTEM LIFETIME

Session Chair – Katherine Edwards (Vodafone)

WE1A – 1 Applying the Hugo upgrade model to re-deployed systems

Phil Lancaster (Vodafone)

WE1A – 2 Tracking the End-of-Life of a Submarine Cable

José Chesnoy (Independent Consultant)

WE1A – 3 Cable Recovery and Redeployment – A Concept to Last?

Alan Proudfoot (Xtera Communications)

WE1A – 4 Mitigation method of spectrum gain deviation in D+ based long distance submarine cable systems with large bandwidth repeaters

Kohei Nakamura (NEC Corporation)

WE1A – 5 Methods and Limits of Wet Plant Tilt Correction to mitigate wet plant aging

Loren Berg (Ciena)

ORAL SESSION WE1B – MARINE ACTIVITIES

Session Chair – Graham Evans (EGS)

WE1B – 1 Shallow Water Cable Abrasion – Managing the Risk

Gordon Lucas (Alcatel-Lucent Submarine Networks)

WE1B – 2 Installing subsea structures – A Successful Cable End Module Case Study

Paul Deslandes (GMSL)

WE1B – 3 The Future of Marine Survey - Applications for Submarine Cables

Ryan Wopschall (Fugro)

WE1B – 4 Technology developments are enabling the

new generation of cable burial ploughs to operate more efficiently with a reduction in operational downtime

Julian Steward (IHC Engineering Business)

WE1B – 5 Big Challenge to Overcome Difference in Bending Radius in Housing Unit Used in Submarine Telecommunication System and Seismic and Tsunami Observation System

Satoki Fujimori (Kokusai Cable Ship Co.)

Wednesday 20th April – Oral Sessions WE2A & WE2B 2.00PM - 3.30pm

ORAL SESSION WE2A – NETWORK TOPOLOGY

Session Chair – Elizabeth Riviera-Hartling (Ciena)

WE2A – 1 Cost allocation on a shared fiber pair using ROADM Bus

Marc-Richard Fortin (GlobeNet)

WE2A – 2 A Solution For Flexible and Highly Connected Submarine Networks

Arnaud Leroy (Alcatel-Lucent Submarine Networks)

WE2A – 3 Integrated Submarine and Terrestrial Network Architectures for Emerging Subsea Cables

Mohan Rao Lingampalli (Equinix)

WE2A – 4 Open Cables and Integration with Terrestrial Networks

Mark Enright (TESubcom)

WE2A – 5 How Resilient is the Global Submarine Cable Network?

Andy Palmer-Felgate (Verizon)

ORAL SESSION WE2B – MARINE ASSETs AND Observatories

Session Chair – Kate Panayotou (GHD)

WE2B – 1 A Paradigm Change to Submarine Telecom Marine Assets

Charles Collins (3U Technologies LLC)

WE2B – 2 Marine Maintenance Synergy's – Sharing costs within new market spaces!

Stephen Holden (GMSL)

WE2B – 3 S-net Project, Cabled Observation Network for Earthquakes and Tsunamis

Toshihiko Kanazawa (NIED)

WE2B – 4 Installation of new seafloor cabled seismic and tsunami observation system using ICT to off-Tohoku region, Japan

Masanao Shinohara (The University of Tokyo)

WE2B – 5 Power and data over fiber for sea-floor observatories

Florent Colas (Ifremer)

Thursday 21st April – Oral Session TH1A & TH1B 10.00am – 11.30pm

ORAL SESSION TH1A – LINE DESIGN

Session Chair – Izumi Yokota (Fujitsu)

TH1A – 1 Capacity limits of submarine cables

Edouardo Mateo (NEC Corporation)

TH1A – 2 Impact of frequency separation between orthogonal idlers on system performance

Pierre Mertz (Infinera)

TH1A – 3 Quasi-Single-Mode Fiber Transmission for Submarine Systems

John Downie (Corning)

TH1A – 4 The Challenge of Very High Cable Capacity: PDM-8QAM modulation format, the booster for submarine cable capacity and $OSNR_{WET}$, the parameter to evaluate cable capabilities

Pascal Pecci (Alcatel-Lucent Submarine Networks)

TH1A – 5 Optical Designs for Greater Power Efficiency

Alexei Pilipetskii (TESubcom)

ORAL SESSION TH1B – PROJECTS AND Impacts

Session Chair – Andy Palmer-Felgate (Verizon)

TH1B – 1 Flexible ROADM Networks: New aspects of Commissioning, Operation and Maintenance through project examples

Jean-Pierre Blondel (Alcatel-Lucent Submarine Networks)

TH1B – 2 The challenges of completing an Oil & Gas Cable System order

Jiawen Wang (Hengtong Marine Cable Systems)

TH1B – 3 Lighting the Way to Bandwidth Equality: The Role of Submarine Connectivity in Bridging the Bandwidth Divide

Michael Ruddy (Terabit Consulting)

TH1B – 4 Evolution of the Internet of the Middle East

Doug Madory (Dyn)

TH1B – 5 Capturing the Public Imagination: Communicating the cultural significance of submarine internet cables

Bronwyn Holloway-Smith (Massey University)

Thursday 21st April – Oral Session TH2A & TH2B 2.00Pm – 3.30pm

ORAL SESSION TH2A – WET AND DRY TECHNOLOGIES

Session Chair – Edwin Muth (TE Subcom)

TH2A – 1 Proving And Qualification Of A Sea-Earthing System For A New-Generation Submerged Branching Unit

Ian Watson (Huawei Marine Networks)

TH2A – 2 Improving the Crush Resistance of Submarine Cables

Weiwei Shen (Hengtong Marine Cable Systems)

TH2A – 3 SLTE Modulation Formats for Long-Haul Transmission

Bruce Nyman (TESubcom)

TH2A – 4 Ultra high capacity transoceanic transmission

Gabriel Charlet (Alcatel-Lucent Submarine Networks)

TH2A – 5 Optimising Design of Dynamic Fiber Optic Riser Cable using Cross Section Analysis

Bjørn Konradsen (Nexans)

ORAL SESSION TH2B – THE LAST SESSION

Session Chair – tbd

TH2B – 1 The future of changing liabilities and effective subsea asset management

Benjamin Sims (Vodafone)

TH2B – 2 Government Surveillance, Hacking, and Network Security: What Can Submarine Cable Operators and Their Customers Do?

Kent Bressie (Harris, Wiltshire & Grannis LLP)

TH2B – 3 Regulation of Underwater Sounds

Richard Hale (EGS)

TH2B – 4 Permanent Reservoir Monitoring (PRM) System Installation: The Installer's Perspective

Andrew Lloyd (GMSL)

POSTER SESSION

Wednesday 20th April – Poster Session 3.30pm – 6.00pm

MARKET & PROJECT – TRENDS & CHALLENGES

MP01 The challenges of fibre optic installation in developing markets

Paul Deslandes (GMSL)

MP02 Submarine Optical Cable Market in Africa: Challenge and potential

Ali Ben Abdallah (Tunisie Telecom)

MP03 How media attention can shorten repair times for damaged cable systems

Kristian Nielsen and Kieran Clark (Submarine Telecoms Forum)

MP04 How EPIC are submarine cable systems?

Ritesh Dass (Vodafone)

MARINE SERVICES AND OPERATIONS

MS01 Submarine Cable Spatial Planning Discussion Based on Increasing Marine Activities

Hongli Shi (Huawei Marine Networks)

MS02 Shore End Cable Protection Case Study – Submarine Cliffs

Zhang Xiaolong (Huawei Marine Networks)

MS03 New Subsea Optical Fibre Junction Box for reduced tensile load applications

Craig Beech (GMSL)

MS04 Lessons Learned from Past Cases for Achieving Disaster Resilient Routing and Economically Viable Maintenance Operation

Yukitoshi Ogasawara (Kokusai Cable Ship Co.)

MS05 Applied seabed geomorphology in cable route planning, surveying and engineering

Elias Tahchi (EGS)

MS06 Real-time Correlation of Cable Fault to Vessel Location and the benefit to Cable System Operators

Darwin Evans (Ciena)

MS07 The Global Challenges of Comprehensive Undersea Jointing

RK Stix (TE Subcom)

NETWORK ARCHITECTURE AND SYSTEM DESIGN

NA01 Power Feeding Solution for Festoon-like Repeatered Submarine Cable System

Li Yuhe (Huawei Marine Networks)

NA02 Spectrum sharing in a multivendor environment

Darwin Evans (Ciena)

NA03 Upgrading on the Longest Legacy Repeatered System with 100G DC-PDM-BPSK

Jianping Li (Huawei Marine Networks)

NA04 Transoceanic Transmission over 11,400km of installed 10G System by Using Commercial Dual-Carrier 100G

Ling Zhao (Huawei Marine Networks)

NA05 Modeling of nonlinear fiber effects in systems using codirectional Raman amplification

Lutz Rapp (Coriant)

NA06 Optimization of Pulse Shaping Scheme and Multiplexing/Demultiplexing Configuration for Ultra-Dense WDM based on mQAM Modulation Format

Inoue Takanori (NEC Corporation)

NA07 Simple method to estimate repeater span by varying the length of a reference digital line section

Francis Charpentier (Orange)

NA08 Resource Savings Using Gridless Submarine Networks based on Filterless Coherent Transmission Technologies

Md. Nooruzzaman (Ecole de technologie supérieure, Université du Québec)

NA09 Benefits of Digital Sub Network Connection Protection for Dual Route Backhaul

Benoit Kowalski (Infinera)

NA10 Innovative Submarine Transmission System using Full-Tunable ROADM Branching Unit

Takehiro Nakano (NEC Corporation)

EQUIPMENT AND COMPONENT TECHNOLOGIES

EC01 The Electrochemical Aspects Of The Use Of Titanium And Steel In Submerged Plant And Cables

Ian Watson (Huawei Marine Networks)

EC02 The Relationship of Carbon Backfill and Metal

Corrosion Rate in Submarine Electrode Array System

Kai Sun (Huawei Marine Networks)

EC03 A more reliable pumps redundancy design

Changwu Xu (Huawei Marine Networks)

EC04 A New Cable Failure Quick Isolation Technique of OADM Branching Unit in Submarine Networks

Hongbo Sun (Huawei Marine Networks)

EC05 Evaluation of Nonlinear Impairment from Narrow-band Unpolarized Idlers in Coherent Transmission on Dispersion-managed Submarine Cable Systems

Masashi Binkai (Mitsubishi Electric Corporation)

EC06 Effective Application of KCS Cable Probe for Localizing Submarine Telecommunication and Power Cables

Takaharu Etou (Kokusai Cable Ship Co.)

EC07 Ultra-low loss and large Aeff Pure-silica core fiber advances

Hideki Yamaguchi (Sumitomo Electric Industries)

EC08 Evaluation of Mixed Metal Oxidation For Sea Earth Electrode With High Reliability

Hong-ying Chao (Huawei Marine Networks)

EC10 Transmission over unrepeatered 85dB fiber link using advanced modulation format

Xiaohui Yang (Infinera)

EC11 Tool-Kit for Ultra-Long / High-Capacity Repeaterless Systems

Philippe Perrier (Xtera)

NETWORK OPERATIONS AND CARRIER SERVICES

NO01 When do pump failures prevent system re-use or lifetime extension?

Tony Frisch (Xtera)

NO02 Real-Time Wet Plant Health Monitoring and Automation

Darwin Evans (Ciena)

NO03 Capacity optimization of submarine cable through smart spectrum engineering

Vincent Letellier (Alcatel-Lucent Submarine Networks)

OIL & GAS AND SPECIAL MARKETS

OG01 Emerging subsea networks: new market opportunities for, and societal contributions from, SMART cable systems

Christopher Barnes (University of Victoria)

OG02 Design of Fiber Optic Cables and Accessories for Offshore Applications

Inge Vintermr (Nexans)

OG03 Fibre to Platform Connectivity – Working in the 500m zone

Andrew Lloyd (GMSL)

OG04 Application of QAM signals to Oil & Gas OADM submarine cable systems

Hiroshi Nakamoto (Fujitsu)

OG05 The European Multidisciplinary Seafloor and water-column Observatory - the Development and utilisation of large scale distributed EU cabled marine research infrastructure

Paul Gaughan (Irish Marine Institute)

SubOptic
2016
www.suboptic.org

Emerging
Subsea
Networks

To see the latest information about
SubOptic 2016 go to our website –
www.suboptic.org

The world's expanding treasure

Dubai

18th-21st April 2016

Celebrating

30

years

of SubOptic

Hosted by

ESTABLISHING A NEW TYPE OF SALES CHANNEL

BY MIKE CONRADI

Abstract:

Since April 2015 DLA Piper have been advising the Africa Coast to Europe (ACE) submarine cable consortium of around 20 West African telecoms operators on an unprecedented and ground-breaking project for the global submarine cable industry, to set up a new form of sales channel for international telecoms capacity.

The ACE consortium (www.ace-submarinecable.com) members have built about 9,000km of the total 17,000km, US$700 million, submarine cable system extending from France to South Africa. The ACE Members, representing around 15 different countries along the route, built and own the cable collectively, and each member uses the capacity it created. However – in common with many other cable systems round the world – the technology used means that the cable has a very large amount of unused spare capacity.

As presently structured, the consortium model used to build the cable means that each member has an allocated share of the capacity which it can only use in respect of landings where it is a "landing party" (meaning that it is given rights under the consortium agreement to offer connections to that location), and there is also a central pool of unallocated capacity. The ACE consortium is not set up to exploit the spare capacity efficiently for sales to third party

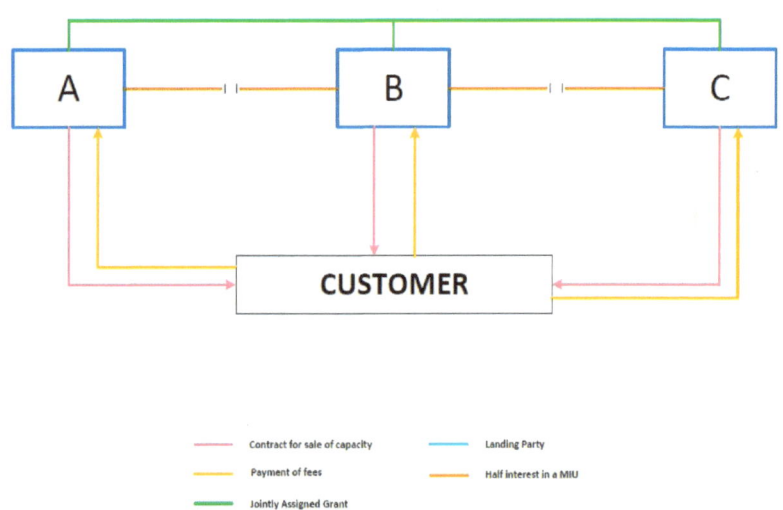

wholesale (or large enterprise) customers – especially where these customers are based overseas. In order to establish a circuit between, say, three points on the cable, a customer must, at present, enter into three different contracts with up to three different ACE Members as shown in the diagram below:

The table above illustrates how a customer wanting connectivity between points A, B and C would need to enter into separate sales contracts with (possibly separate) landing parties at each of those locations, and that those landing parties would, in turn, need to arrange for part of their half-interest in "MIU"s (minimum investment unit - the unit used for calculation of investment into the consortium) under the consortium agreement (the "C&MA") to be assigned for the use of that customer between the relevant locations.

This is clearly inefficient and results in a great many missed sales opportunities for the ACE Members. The other consequence of the consortium structure is that many of the landing parties, being smaller-scale telecoms operators, have only limited capability to market and sell their capacity to large-scale international wholesale customers, like large traditional telecoms operators, or else to large multinational enterprises (like oil companies) or Over The Top (OTT) businesses like Google and Facebook. As a result they are losing opportunities to generate additional revenues.

Solution

DLA Piper has been helping design and structure a new sales channel – ACE Connect – which will have rights to take and re-sell spare capacity on the ACE cable.

The structure involves establishing a new company – ACE Connect – as an SPV (special purpose vehicle) to be owned by all those ACE consortium members that wish to participate (not necessarily all of them). It will be established and structured in such a way as to minimise and simplify its tax obligations and will employ (or engage) specialised sales staff whose role will be to sell spare capacity on ACE, focussing especially on international wholesale and large enterprise customers.

ACE Connect will enter into a framework agreement with each of its shareholders under which they each agree to offer to set aside capacity to ACE Connect on pre-agreed terms as and when required for a customer opportunity. The agreement also deals with a complex allocation of any revenues as between all the participating parties - with all participants receiving a portion of revenues whether or not any particular sale uses "their" capacity (or goes via landing stations for which they are themselves a landing party). The ACE consortium may also decide to amend the C&MA so as to facilitate ACE Connect's operations in various ways.

Other aspects of the project have included drafting the shareholders' agreement (between those members of ACE that which to participate in the establishment of ACE Connect) - dealing with issues like how members can join and exit the arrangements as well as how ACE Connect'sboard will be composed and decisions made about it. ACE has also had to analyse the tax effect of various possible cashflows through this model.

This is a highly innovative project – although there are one or two structures with some similarities in the industry there is nothing set up in the same way or with the same aims. ACE Connect will, following its establishment, allow participating operators to have access to an entirely new sales channel, thus facilitating their own business as well as acting for a model for further telecoms infrastructure investment in emerging markets.

DLA Piper won the TMT Law Firm of the Year awards at the African Legal Awards for our work on the project.

Mike Conradi is one of the lead partners for telecoms matters at DLA Piper, which is one of the world's largest law firms. He is ranked as one of the leading telecoms lawyers globally by the various legal guides, with Chambers & Partners describing his "ability to grasp complex technical points quickly" and as well as his "skill in navigating through the constraints of a tough regulatory environment".

Mike has particular expertise in the submarine cable sector. He has worked, to varying degrees, on more than 30 different systems and was the only private practice lawyer to sit on the Legal working group for SubOptic, which drafted a template system supply agreement.

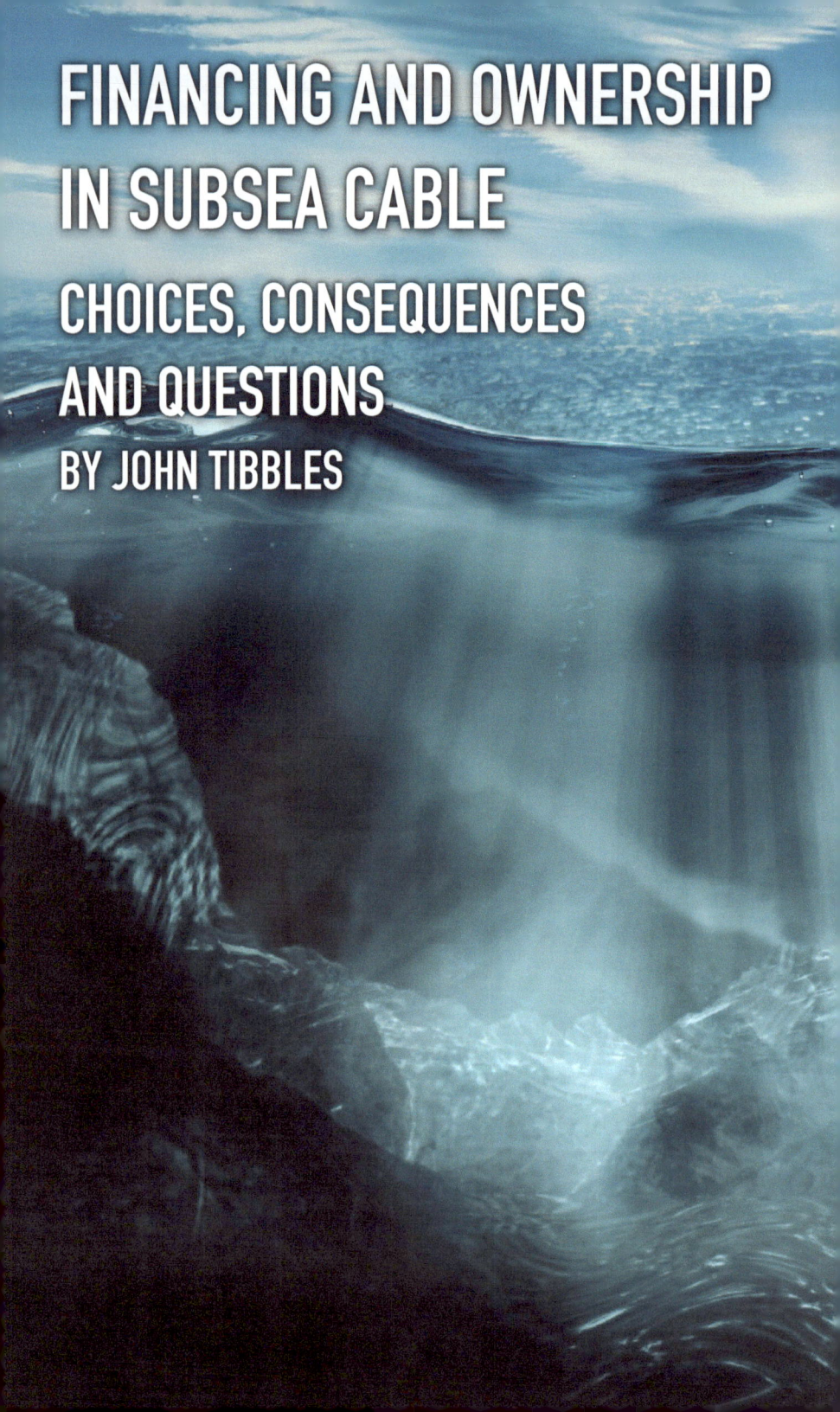

Abstract:

There are several common methods of financing a submarine cable, each one presenting its own difficulties regarding profitability. This article examines the different 'financing fashions' and some speculation about where they have lead us and what may happen next.

Building submarine cables has always been a risky business. That is why consortia were developed to mitigate risk on what have always been expensive projects. In the 1980s and 90s, so called private systems appeared alongside consortia but the risk hadn't gone away. The private ventures were by and large unsuccessful financially, spectacularly so in some cases resulting in some serious distortion to the market leading Atlantic route producing an artificially low cost base which destroyed the value in better funded and seemingly risk proofed projects. If that wasn't bad enough, technology in the form DWDM and faster basic line rates came along to increase the capacity of these systems by a factor of up to 10 fold. The consequential dramatic reductions in the price of capacity and thus underlying investment value to virtually nil while at the same time this cheap capacity allowed cloud computing and OTT services to develop and flourish.

This doesn't create a positive scenario for building new systems to keep up with the seemingly ever increasing demand to watch cats falling off fences. With the oversupply of capacity finally showing signs of exhaustion in the last few years new systems were finally needed but who decided which projects would work and which wouldn't. A question that was ever thus but now it isn't the traditional mega carriers deciding the fate of entrepreneurial cables but big data /OTT or as I call them – Neo carriers. While this seems to be good news for them, because they now set the agenda on construction and capacity cost, it brings with it some potential complexities for their overall businesses and subsea cable industry.

A couple of issues back I wrote a piece called "the rise of the 'Neo Carriers" and how they were supplanting the traditional big western telecoms carriers. Well, the trend has continued apace at least in the West, but to the point where it starts to raise some tricky questions for them and for the industry. However, as so often is the case, the story needs to start in the past with how cable financing and its implications developed and how we got to where we are before looking further forward.

How Cable systems are Financed and Why is it Important.

The way in which a system is financed isn't just about raising money, it dictates the ownership and management structure and pricing strategies during the build and throughout its operational life

Coaxial Consortia

This short history starts in the grey days of the 1950s,

submarine cables of course had been around for decades serving the worlds telegraph companies. World War 2 brought about a revolution in demand for secure intercontinental voice traffic and the development of coaxial cables for subsea use made that possible and along with it intercontinental phone calls that did not hiss and crackle and disappear altogether with the vagaries of weather and the dawn and dusk of every day.

The first true intercontinental system was TAT1 which linked the USA and Western Europe, the partners were the state carriers, administrations, PTTs etc. as they were known, of USA and Canada and Britain and France with invited participation from some favored other European countries to form what was effectively the first cable consortium and the relevance to this article is just what was a consortium and why did it make sense,

There were three main drivers for this:

1. Risk mitigation: this was new and very expensive enterprise with new technology and carried considerable risk.

2. Political issues: this is not that long since WW2 and the cold war is very much in evidence, no country was going to let another, even a friendly one operate strategic communications facilities on its territory and outside the USA, at that time a virtual monopoly in ATT, telecoms operators were all state owned monopolies.

3. Economic structure: At that time almost the demand for cable capacity was for voice with around ten percent of capacity for leased lines and bearer circuits for services like telex. In all of these cases a revenue sharing and cost sharing scheme applied with each sovereign party sharing half the costs of the link and getting a share of the revenue, in the case of voice via the accounting rate

system which shared revenues between the caller and called countries.

Overall, this concept shared the construction and development risk along with capital and operating costs and the system management structure and carriers consequently shared the reward for the use of the systems for voice and telex etc.

TAT1 was full within weeks of going into service and was rapidly followed by a succession of other transatlantic systems (TAT) with larger capacities and in the Pacific similar technology systems were built. All were constructed upon the same share risk share reward model which forms the basis for what we know as the Consortium cable and this model and technology continued successfully through to the 1980s and TAT7.

Fibre Optics and Freedom?

Coaxial technology was always behind the demand curve, the cost of a single circuit on thousands of kilometres of cable when such a system could only carry at best around four hundred 4Khz voice channels was prohibitively high and cables although important became very much second choice to the higher capacity and far more flexible and

lower cost Intelsat system which dominated international communications in the 1970s and 80s. Intelsat although heavily American influenced was a truly globally run and managed project and served the world very well for over 15 years, but in 1988 a new cable appeared, this was TAT 8. TAT 8 was different, -it was fibre optic and it was digital and as all carriers knew from their domestic networks that digital telecoms was the future.

In addition to technology making TAT 8 different there had been changes at the political level in telecommunications, the USA had broken the ATT Bell System monopoly and European countries like UK and Sweden were licensing new carriers in the form of Mercury and Tele2 to compete with the state owned but soon to be privatized giants and this trend spread across Europe.

TAT8 was built on the consortium model, but this time the consortium process had another dimension added to it, an anti-competitive one

since consortia consisted of invited members and the new 'private' carriers were not invited or faced other barriers to entry. One consequence was a surge of regulatory protests but also, in 1989, the first private cable for decades in the form of the Anglo American (Cable and Wireless: Sprint) PTAT cable which competed head on with the TAT8/9 systems.

So now, for the first time since the telegraph age, there were real competing systems with competing business models. The consortium required payments matched to the system build costs and delivered capacity 'at cost' in proportion to ownership. The private cables aimed to make a profit. They didn't charge customers until the system was ready for service and let anyone purchase capacity who was duly licensed whereas the big 'club' carriers seemingly wanted to starve their new competitors of the network capacity vital for new entrants to get their business going.

The other huge difference between the traditional consortium and the 'private system' was that the latter had a large element of the financing in the form of debt from external investors and banks. This was structured in different ways in different systems by the financial professional but for the sake of simplicity let us just refer to it as debt financing whereby system promoters put in equity, some presales aided cash flow and the assurance of demand and the finance community provided debt financing in a similar way as they would with any large commercial venture. As the new century dawned they found to their cost that submarine cables are not just any commercial venture. This all leads nicely on to the troubled times of the Dotcom bust.

That word Risk again

Debt financed private cables were for a while all the rage and seemed to be in many ways, except the important element of liquidity, superior to consortia. They were relatively agile in decision making, did not have rotating Chairman and lots of committees to manage build and operations and it seemed they could be successful in selling in competition against consortia systems. In practice though many of these sales were because there was no alternative direct path or the private system was needed for diversity, or was over exposed to underfunded new carrier ventures . So, when demand for capacity dried up cable sales dried up too and as is well known many systems suffered severe financial problems ranging from refinancing on more onerous terms to outright bankruptcy. It was this latter development that caused a massive glut of supply as the systems either had fire salesj to stave off bankruptcy or over the longer term were bought out of bankruptcy for a fraction of their original cost giving their new owners an unassailable cost base. Prices collapsed, even the fully funded consortia could not compete and could only follow the spiral and see their value destroyed. For a long spell development of new systems almost completely stalled except where there was a desperate need for a new system as was the case in the Indian Ocean, the answer to that problem was SMW4 and it was very very much a consortium project. As for the financial community it was clear that submarine cables were not for them.

Winding the clock forward to around 2011 and after many years where getting any finance for a new build , even internally financed consortia , was close to impossible demand pressures on parts of the global network finally began to lead to some new development. Mostly in Africa and the Indian Ocean region where there was just not enough, or in some cases any, capacity and gradually new systems began to appear. These were mostly consortia or complex variations thereof like EAS-Sy formed because no one

party would take on the responsibility for risk of system development. This era also marked the emergence of cables as something carriers actually needed, whether their finance directors liked it or not, the reason was the real, much delayed, emergence of high volume broadband internet, the first steps to global LANs and the emergence of cloud computing. These developments finally produced the levels of demand predicted back in the early years of this century and which had led to the oversupply situation as new systems were built on the promise of growth that didn't materialize for ten years. Now there were huge corporations with huge needs for data transport such as Google and of course there was one more factor that was new and that was the emergence of China as a massive source of demand for subsea cable capacity.

What the world needs is Unity

And, the first step into this new world was, appropriately enough lead by Google, who after deciding there was little point in paying a carrier middle man for their capacity had begun to buy up wavelengths in the Atlantic taking full advantage of the artificially low prices. But in the Pacific this option didn't exist as the oversupply was on a much more modest scale, in the Pacific a new system was needed for Google to achieve their objectives.

However, there was no new cable on the horizon, the traditional US giants of ATT/ Verizon were now run by managers from the mobile side of their business who knew the checkered history of cable investments and were not inclined to follow that path. Equally an un-wieldy consortium was certainly not what Google (or more precisely those running their small subsea cable teams) had first-hand experience of the problems of both consortia and private systems. They realized something new was required. Google had perhaps an unlikely ally in this in China Telecom but allies they were and a new cable was born called Unity.

Unity was new because although it was a consortium it had a dramatically new element to it-to join you had to commit to a whole fibre pair-a very big investment in subsea cable terms and

an unprecedented level of commitment and it worked. Unity was built and is working today and has become a benchmark for a new model of financing cables often referred to as a condominium approach. Although technically a consortium it is very different in practice with a much streamlined management process and the right for individual fibre pair owners to do much as they wished with what in effect was their own single pair cable. Something of extreme importance when it comes to upgrades an area fraught with problems in traditional consortium systems.

Low Latency Side Show

During this period there also emerged one more new potential demand driver that prompted another ground breaking development-this was latency or rather lack of it. Achieving a low latency on a cable system has always been important but never enough to actually dictate the route and engineering in the face of known obstacles. However, the demand for high frequency trading between financial centers and its complex processes established a market for a system that had the lowest overall latency as this could be translated into competi-

tive advantage for the traders. One company, Hibernia, came up with a design to serve this market and called it Hibernia Express. But although it was low latency that gave them the idea, on its own it was not enough to sell substantial capacity at a premium price for the lower latency to make it work in the fiercely competitive trans-Atlantic market. However, after several long years trying the Neos, OTT and Big Data, companies realized that unlike most of the older Atlantic systems Hibernia could sell them a whole discrete fibre pair. Sales on that scale assured the financing and Hibernia Express is now in service funded by big data not the low latency but low volume of the financial trading world.

New Questions for New Faces

In the few years since Google started Unity, Big Data, Neo carriers, OTTs, call them what you will, have become very involved in the submarine cable arena, global gi-

ants such as Amazon, Facebook and Microsoft, looking to buy capacity at the fibre pair level. The large traditional Asian carriers, CT, NTT, KT, Telstra, and Sing-Tel etc. have continued to invest heavily in cable systems and have included if not embraced the Neos as well funded co owners. The picture in the West though has been very different. The huge oversupply in the North Atlantic, compounded by the new upgrade technologies, has meant only two new cables have followed the last of the older systems which was Apollo completed in 2003. Both of these, Hibernia Express and the Aqua project (formerly Emerald)

are now in service and that would not have happened without fibre pair purchases from the Neo Carriers. The same can be said of systems like Monet (US-S. America) which would be unlikely to be in business without Google's participation.

Of course it is only natural that the Neos drive a very hard bargain for their fibre pair level purchase, essentially aiming to get capacity very close to cost. What 'at cost' actually means is of course open to many interpretations, but a sale to just one such party can make all the difference between getting finance for a project or it failing.

And this begs some very important questions-

With these large scale sales/ investments at low margins what is really left to sell on the cable-the owners face a hard road in selling to other users at much higher margins than the Neo carrier customer has secured. Multiple sales mean almost the whole cable is owned by the new breed of customer leaving little scope for any profit at all. So where is the incentive in such a model for a new system since the margins have been compromised to make the initial key sales and upgrade rights are sold along with each fibre pair?

With their emerging new super networks, the Neos realize they need redundancy and diversity for their transport and cloud type offerings but how do you load share or back up a fibre pair, well you do it with another pair on another system-also bought 'at cost' or you split the pair into 'spectrum' and trade half the fibre pair with another Neo who bought a pair on another system.

Initially, it seemed, the Neos didn't cooperate because at the highest corporate level they were competitors-and indeed still are, but it makes financial sense to cooperate on cable capacity. Indeed these competitive dynamics are no different and rivalries no worse than those between ATT and MCI WorldCom or BT and Cable and Wireless a decade and more ago. The difference is that a decade and more ago these big collaborating competitors were not Neo carriers they were Carriers and very heavily regulated ones at that when it came to subsea cable systems. Phrases like dominance and market power and open access were in daily use and any cable project they sponsored was subject to great regulatory scrutiny.

We now have a situation where the Neo carriers dominate the present and foreseeable future of subsea cable systems in the Atlantic region and heavily influence them elsewhere. They don't build cables yet but the comparatively tiny cable development and operating companies are completely beholden to them financially. After selling at very very low margins on a large scale to key investors there is limited scope for making substantial profits. Consequently, they could go the same way as some of the earlier private systems and collapse when lack of future revenues meant they could not meet ongoing Opex demands like O&M expenses. Some have suggested that the Neos (well some of them) have had this as a plan all along, invest in a minnow company with a big purchase , wait for it to file Chapter 11 and then pick it up for 15c on the Dollar . Maybe? Maybe not?

While this rather cynical ploy might make financial sense it seems to my mind that they would not do this. Most of them are beginning to encounter difficult regulatory relationships on a bigger scale than just telecoms regulators but national and supra-national anti-competition bodies and avoiding or minimising trouble there seems to me to be worth much more than some cheap cable capacity.

Equally when the need for another Atlantic cable comes along maybe the Neos will go that small step further in co-operation and build it themselves, Gemini style, since no one else has the appetite or money. Indeed, something like this is already the subject of informed speculation. But if they do go down that route they are no longer neo carriers but carriers pure and simple. –I doubt even Google's resources could spin the idea that owning numerous local networks, long distance capacity in the USA and controlling Subsea cable builds as well could somehow mean they are not a de facto carrier and therefore should be regulated as such. They are not alone in this situation since Facebook and Microsoft and the others face all or part of the same dilemma. It seems unlikely that they would want to put themselves in the position of being classed as carriers and subject to the often onerous regulation that comes with it but if no one else will build or fiancé new cables what are they to do.

So, a few more questions need to be asked and at some point answered. Among the questions, in my opinion are: -

1. Do the Neos exhibit excessive market power in the subsea segment

2. Will they try and spin off cable ownership and management into subsidiaries that if necessary can be regulated without it impacting on their diverse main businesses

3. Will they be comfortable that their multibillion dollar enterprises and service delivery functions depend on companies not even one hundredth of their size

4. On the other side of this coin will the former titans lose interest in cable to such a degree that they will seek a 'cell tower solution; selling off assets and leasing back what they need via venture funds or specialized asset management companies. Perhaps no longer being directly involved in con-

struction and operation of assets that for years were at the core of their business.

5. Will the EU be concerned that on their most important trade path, the Atlantic all cables and all that capacity is controlled almost entirely by American companies with all the questions that raises both in a telecoms context and wider political ones

And Finally

As time passes some things change and some things don't, subsea cable projects remain challenging investments which ask complex questions of those investing in them or purchasing capacity on them. In a few short years the subsea capacity and new build market has very radically changed with the titans of the past disappearing from the scene while in the East the big traditional players have made absolutely sure that while they welcome investment from the

OTTs they are not about to relinquish control of a strategic asset any time soon. In the West however the free market is still there and anyone can try to build a subsea cable.

However only a very few entities have the power to choose where to invest and thereby make or break a project. Will that fact put the regulatory spotlight on a whole new group of people? Google 'dominance' to find out? Or perhaps click on Amazon.com to buy your network capacity on line. After checking out the Facebook page of someone with capacity to sell of course.

John Tibbles has spent over 30 years managing globally based investments in cable systems for some of the worlds major subsea network operators and owners involving strategic planning, partnerships and consortia management , buying and selling in the wholesale space and managing supplier relationships. He has been actively involved as a panelist, presenter and member of many industry bodies including SubOptic, PTC, ICPC as well as contributing to media articles on the industry. Now retired from daily involvement he owns JTIC consulting (www.consultjtic.com) providing consulting services for the submarine cable sector and the broader international carrier business

The Power of Submarine Information Transmission

There's a new power under ocean uniting the world in a whole new way. With unparalleled development expertise and outstanding technology, Huawei Marine is revolutionizing trans-ocean communications with a new generation of repeaters and highly reliable submarine cable systems that offer greater transmission capacity, longer transmission distances and faster response to customer needs. Huawei Marine: connecting the world one ocean at a time.

HUAWEI MARINE
N E T W O R K S

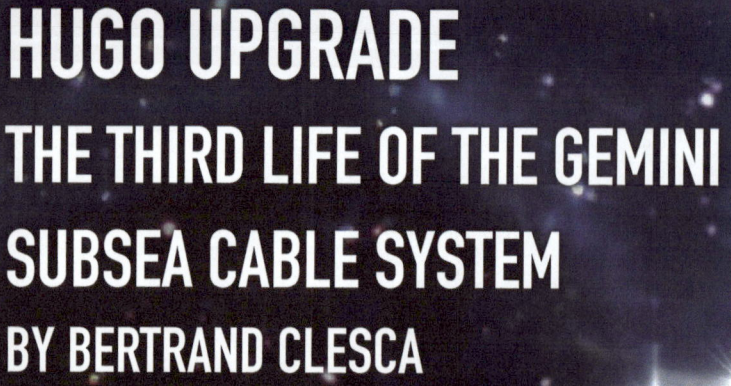

HUGO UPGRADE

THE THIRD LIFE OF THE GEMINI

SUBSEA CABLE SYSTEM

BY BERTRAND CLESCA

Abstract:

There are several common techniques for upgrading a submarine cable system, including upgrading line terminal equipment and recovering and repairing a repeatered system. This article discusses the first upgrade done by the insertion of innovative new repeaters into an unrepeatered system.

Submarine cable systems form the invisible backbone that carries almost all international data traffic. There has been exponential growth in capacity demand fueled not only by increasing number of users but also increasing methods of access and increasing numbers of services. Businesses also now store less data local and rely on cloud services and data centers adding to demand.

One way to extend the lifetime of existing submarine cable systems is by upgrading the line

terminal equipment to higher transmission rates thereby increasing capacity and in the main lowering the cost per transported bit. Such terminal upgrades have become popular since the beginning of the 2000's. Additionally, the use of non-base system vendor equipment for upgrades on dark unused fiber pairs and as shared spectrum on working fibers is now an established upgrade practice.

The recovery and re-lay of out of service submarine cable systems is another way to extend

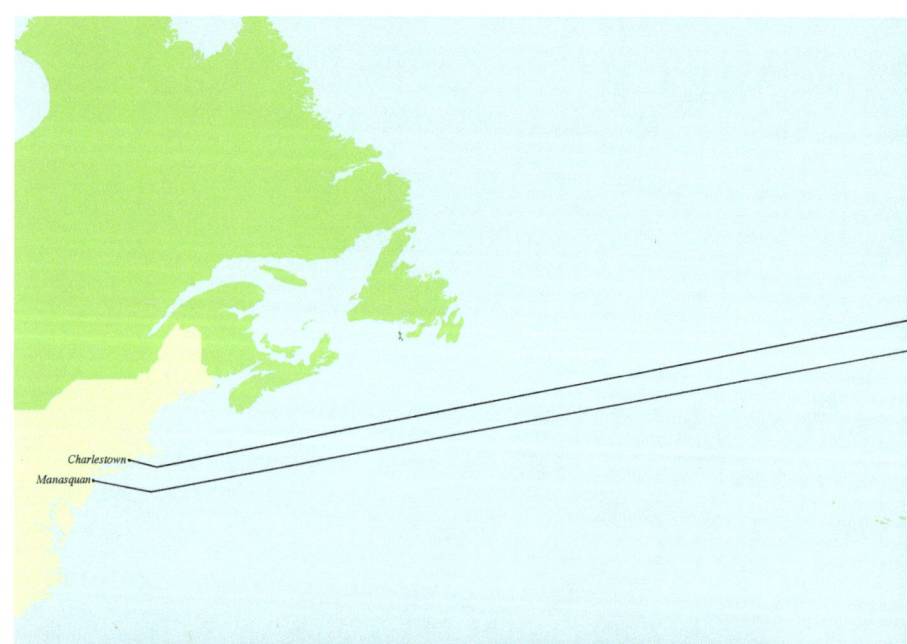

Figure 1: Gemini South and North legs across the Atlantic Ocean

their lifetime. The recovery of a repeatered cable system gives the opportunity to repair faulty repeaters or upgrade the older receive, re-time and retransmit (3R) repeaters (regenerator type) by replacing them with advanced optically amplified repeaters with better optical performance.

This article discusses the first upgrade done by the insertion of innovative new repeaters into an unrepeatered system. The wet plant upgrade of the HUGO cable system increased the capacity and system life of

a cable system that was itself built using re-deployed sections from the de-commissioned Gemini system.

Previous Lives of Gemini Cable System

First Ages

Gemini was a transatlantic cable system connecting Porthcurno (UK) to Manasquan (USA), deployed by Cable & Wireless (now Vodafone) and put into service in 1998. The system used a slotted core cable with LS fibers, with a dispersion wavelength of about 1570 nm. The cable system used early second-generation amplified repeaters with a narrow optical spectrum, enabling an original design system capacity of 4 x 2.5 Gbit/s.

Repeatered Revival

The Gemini system was retired in 2006 due to commercial obsolescence, as newer cables had a lower cost per bit. The Gemini cable was broken up into several sections for re-use by Cable & Wireless. Three of these sections were recovered and re-laid by C&W and Xtera to build regional cable systems in the West Atlantic Ocean. The Gemini Bermuda cable system between M a n a s q u a n and Bermuda (1,572 km), the Caribbean Bermuda US (CBUS) cable system between Bermuda and Tortola in the British Virgin Islands (1,692 km), and East West Cable system connecting J a m a i c a , D o m i n i c a n Republic and the British Virgin Islands (1,745 km). These redeployed cable systems made use of the original Gemini repeaters and were upgraded to transport up to 720 Gbit/s using the most up-to-date transmission equipment at the time of the redeployment.

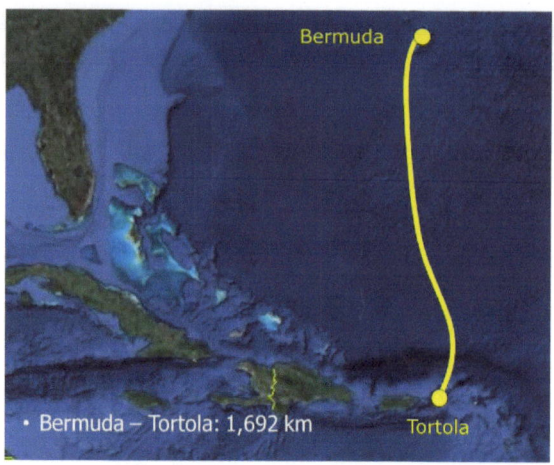

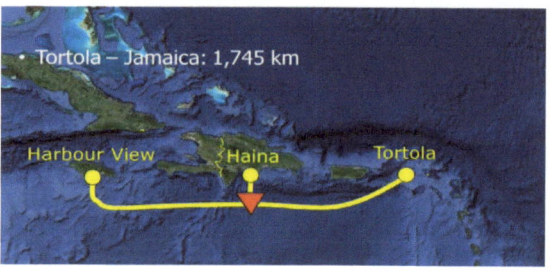

Figure 2: Three regional subsea cable systems built with pieces from decommissioned Gemini cable system.

Conversion to Unrepeatered – HUGO

The eastern portion of Gemini South was re-routed by C&W and France Telecom Marine (now Orange Marine) from Porthcurno in Cornwall to Guernsey (270 km) and from Guernsey to Lannion in France (170 km) in 2006. The repeaters were removed from both segments and returned to the original supplier Alcatel for recycling. The system was called the High capacity, Undersea Guernsey Optical-fiber (HUGO), although the acronym was partly chosen in honor of the author Victor Hugo who lived in Guernsey. Xtera supplied the submarine line terminal equipment to launch 10G

optical wavelengths into both segments. Due to the high zero dispersion wavelength at about 1570 nm, optical wavelengths were allocated above 1590 nm (about 40 waves could be used between 1593 and 1618 nm). The driver for this wavelength allocation was:

- Getting away from the zero dispersion wavelength in order to avoid significant nonlinear Four-Wave Mixing (FWM) effect favored by low chromatic dispersion and high channel power (as required in long unrepeatered link);

- Operating in the positive chromatic dispersion regime for better transmis-

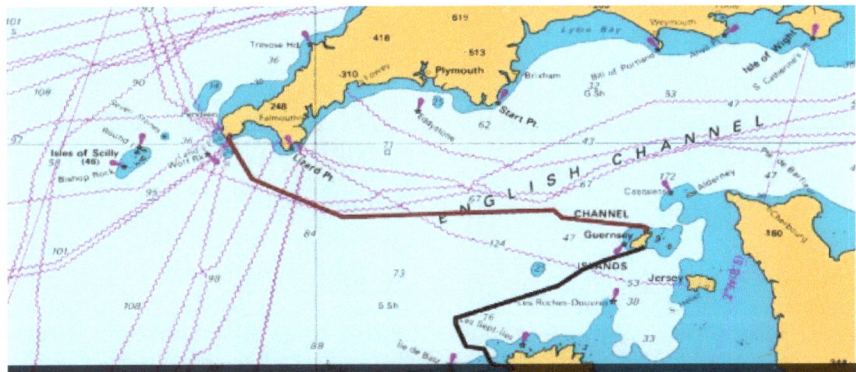

Figure 3: High capacity, Undersea Guernsey Optical-fiber (HUGO) subsea cable system.

sion performance and easier chromatic dispersion compensation.

The English Channel is one of the world's busiest seaways as well as being a fertile fishing ground. As a result, knocks from anchors and fishing nets caused higher than expected losses. Individual losses tended to be small but the large number of them and their distribution along the cable prevented economic repair. This limited the system from reaching its maximum upgrade design capacity with satisfactory performance. Furthermore, the increase in cable attenuation prevented upgrading to 100G waves.

A New Lease on Life: Back to Repeatered!

Two options were considered for increasing the capacity of the HUGO cable system. The first option was the insertion of two Remote Optically Pumped Amplifiers (ROPAs) into the cable. A ROPA is a very simple sub-system that is typically placed 60 to 150 km ahead of the receive end, depending on the system length and fiber attenuation. This sub-system is made on a few passive optical components that are placed inside an enclosure and jointed into the cable. As the ROPA is passive it requires no remote electrical power feeding from the cable end. The energy, necessary for creating optical amplification, is sent to the ROPA by optical pump waves launched into the line fiber from the terminal equipment. Although usually effective, in this case ROPAs would only have enabled a limited increase in capacity, due to the nature of the losses, thus not justifying the cost.

The second option was the insertion of optical repeaters, with the clear benefit to significantly boost the system capacity and optical margin, and enabling an upgrade to 100G waves. This option was possible due to the copper conductor in the original Gemini repeatered cable enabling electrical power feeding. Due to the short segment length only two repeaters were required in

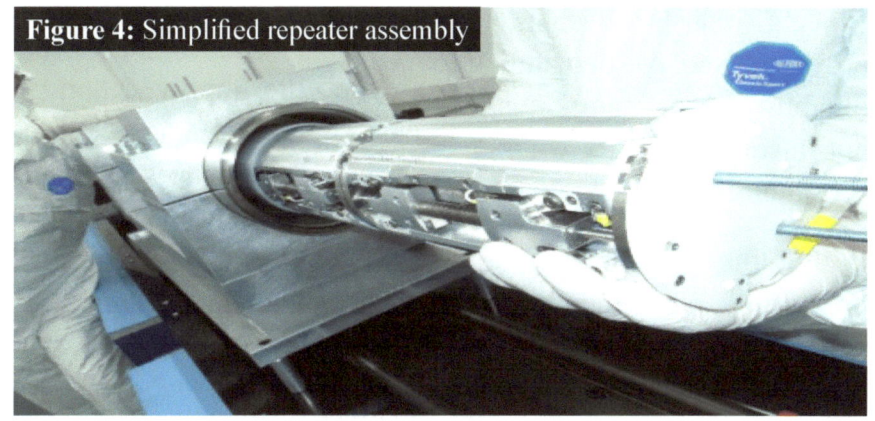

Figure 4: Simplified repeater assembly

the longest segment between Porthcurno and Guernsey making it possible to use a low voltage Power Feed Equipment (PFE).

The return on investment clearly favored the repeater option in terms of capacity, optical margin, ultimate upgrade capacity and system life.

Innovative Repeaters for Creative Undersea Repeater Upgrade

Since the introduction of fiber optics to submarine telecommunications in the 1980's there has only been one major revolutionary change in repeater technology, this being the switch from regenerative (3R) repeaters to optically-amplified repeaters. Other changes have been evolutionary increments of established designs. Such low speed of innovation seems surprising if compared with other advances in undersea communication technology and the tremendous growth in capacity across the globe. The change to optically amplified repeaters made the undersea plant agnostic (to some extent) to data speeds (2.5G, 10G, 100G, etc.). This allowed suppliers to develop ever more advanced transmission equipment to service capacity demands without addressing the limitations of repeaters.

With this in mind Xtera launched a new repeater in 2013. It had to be an evolutionary increment to existing industry designs in order to satisfy demanding reliability requirements, but also needed to offer a substantial capacity increase.

With the advantage of not being restricted by an existing design, Xtera have been able to find modern solutions to old problems. All previous designs of repeaters used Erbium Doped Fiber Amplifiers (EDFA). Due to the physics of amplification, the bandwidth is limited to about to around 35 nm. By combining this amplifier with a Raman amplifier (named after the Nobel Prize winning physicist C.V. Raman), a bandwidth of up to 60 nm can

be achieved.

Beyond the improved optical performance, Xtera's repeater offers the industry's smallest form factor repeater product, enabling easy pass through cable ploughs and facilitating deployment operations.

Project Implementation

The challenges encountered during the system design phase of the new repeatered system are somewhat similar to those faced when unrepeatered line terminal equipment was selected in 2006. The small fiber core, relatively high loss, some ageing and a high zero dispersion wavelength above the conventional

Figure 5: Loading HUGO repeaters on Orange Marine's Pierre de Fermat cable ship.

operation band of a traditional EDFA (the so called C band, typically ranging from 1530 to 1565 nm), were all challenging. In order to limit the amount of fiber nonlinearities, an unusual bandwidth window (above the C band on a wavelength scale) and a lower than usual output power was selected.

The challenging time schedule for the HUGO upgrade project was imposed by a combination of declining optical margin, customer demand for increased capacity, the limited weather window in the English Channel, vessel availability and customers request to minimize down-time. The contract was signed in December 2014, and the system was operational mid-June 2015.

Discussion

Once the repeaters had been installed and terminals upgraded the system performed far in excess of initial expectations. The optical margin is now greater than when the system was originally installed in 2006.

The maximum upgrade capacity is in excess of 9 Tbit/s, which is 225 times that of the original Gemini system. The Gemini cable was first installed in 1998 and after this upgrade can now be expected to carrying traffic until at least 2040.

This type of upgrade has many advantages over a new turnkey build. Not least there is a significant cost advantage and it makes good sense to get the maximum possible life out of an existing investment. However, such an upgrade is not as simple as a turnkey project. There will be unexpected issues that may cause delays and extra costs, which require careful management in order to ensure success. Xtera is grateful to the submarine engineering team in Vodafone and their consortium partners Sure Telecom for helping make this project a success and adopting this new approach to growing capacity demands.

Organizations involved in the upgrade

Xtera Communications, Inc. (NASDAQ: XCOM) have been upgrading submarine systems since 2004 and with its new repeater offering has become the world's most innovative of turnkey submarine supplier.

Orange Marine, previously France Telecom Marine, have been a leading supplier of marine services to the telecom industry for decades. It has recently taken delivery of the one of the industry's most advanced cable ships the Pierre de Fermat.

Sure, originally The States Telephone department, was founded in 1892 and is the leading provider of telecom services in Guernsey and well as Jersey and the Isle of Man.

Vodafone, previously Cable & wireless, has been involved in submarine engineering since its beginning in the 1860's and is still a leading provider of engineering services.

Stuart Barnes joined Xtera in 2007 and serves as the Senior Vice President and General Manager, Xtera Submarine Business. Stuart has over 30 years of experience in the submarine telecommunications business. Prior to Xtera, Stuart was the founder and COO of Polariq, as well as founder and CTO of both Azea Networks and of ilotron. In addition, Dr. Barnes has held senior management positions at Atlas Venture, Alcatel Recherche, STC Submarine Systems and STC Cables Newport. Stuart holds over 20 patents, has published over 40 papers, and has been recently appointed to the Advisory Board of the Aston University Institute of Photonics.

Bertrand Clesca is Head of Global Marketing for Xtera and is based in Paris, France. Bertrand has over twenty five years of experience in the optical telecommunications industry, having held a number of research, engineering, marketing and sales positions in both small and large organizations. Bertrand Clesca holds an MSC in Physics and Optical Engineering from Institut d'Optique Graduate School, Orsay (France), an MSC in Telecommunications from Telecom ParisTech (fka Ecole Nationale Supérieure des Télécommunications), Paris (France), and an MBA from Sciences Po (aka Institut d'Etudes Politiques), Paris (France).

Tony Frisch started at BT's Research labs and then moved to Alcatel Australia, becoming involved in testing submarine systems. A move to Bell Labs gave him experience in terminal design and troubleshooting, after which he went back to Alcatel France, where he worked in Alcatel Submarine Networks' Technical Sales before moving to head Product Marketing. He is now SVP, Repeaters and Branching Unit for Xtera Communications.

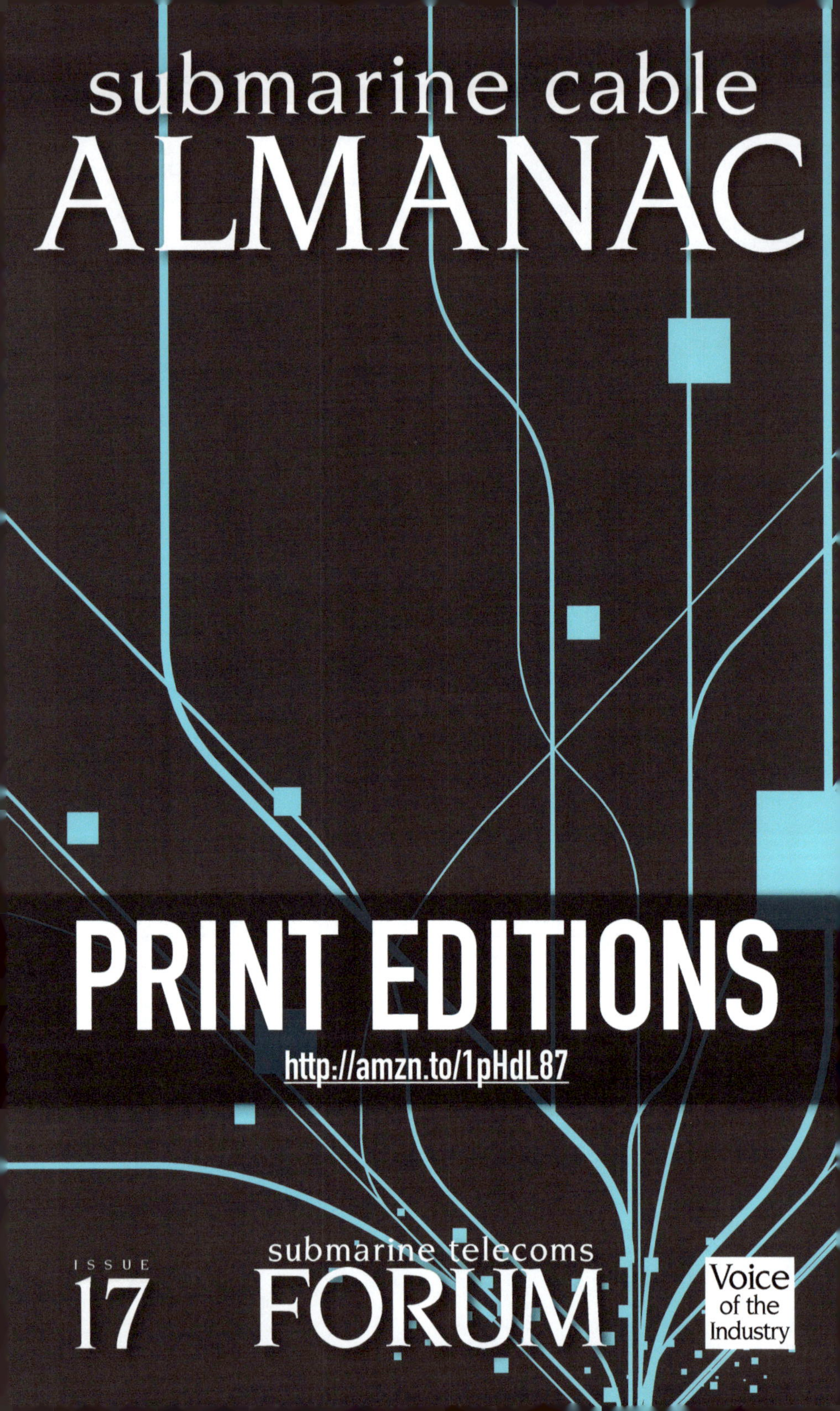

submarine cable
ALMANAC

PRINT EDITIONS

ISSUE

17

submarine telecoms
FORUM

Voice
of the
Industry

submarine telecoms
FORUM

wfnstrategies

celebrating

15

years of excellence

Telecoms consulting of submarine cable systems for regional and trans-oceanic applications

ADVERTISER'S CORNER
BY KRISTIAN NIELSEN

Has it really been three years already? It feels like it was just yesterday that we were drinking wine on The Seine watching the Eiffel Tower sparkle after sunset!

But here we are, a blink of an eye later and SubOptic is upon us again, time again for our industry to come together to exchange ideas, educate, and celebrate the victories of the past three years. A few short weeks from now, the SubOptic conference will be celebrating 30 years of collaboration and education in the submarine industry.

SubTel Forum has its own auspicious milestone this year, we are celebrating 15 years as the voice of the industry. Founded with similar ideals, SubTel Forum is proud to serve this industry alongside SubOptic.

But back to business – we will be producing a daily video wrap-up during each day of the conference. Each day we will be bubbling about, taking interviews and covering presentations, and will later publish a comprehensive wrap-up of the day's activities to our YouTube channel and news feed.

As with all of our products, the wrap-up will be made free to our readers and supported by sponsorships. If you would like to have your company featured as a sponsor of the daily wrap-up, please contact me directly.

Finally, I would like to give all of our sponsors a special Thank You for you stalwart support over the years – without you, there would be no voice of the industry.

See you in Dubai!

Kristian Nielsen literally grew up in the business since his first 'romp' on a BTM cableship in Southampton at age 5. He has been with Submarine Telecoms Forum for a little over 6 years; he is the originator of many products, such as the Submarine Cable Map, STF Today Live Video Stream, and the STF Cable Database. In 2013, Kristian was appointed Vice President and is now responsible for the vision, sales, and over-all direction and sales of SubTel Forum.

 +1 703.444.0845

 knielsen@subtelforum.com

Submarine Telecoms Forum, Inc.
21495 Ridgetop Circle, Suite 201
Sterling, Virginia 20166, USA
ISSN No. 1948-3031

PUBLISHER:
Wayne Nielsen
VICE PRESIDENT:
Kristian Nielsen
MANAGING EDITOR:
Kevin G. Summers

CONTRIBUTING AUTHORS:
Kieran Clark, Bertrand Clesca,
Mike Conradi, Catherine Kuersten,
Andrew D. Lipman, Ulises R. Pin,
John Tibbles

Contributions are welcomed. Please forward to the Managing Editor at editor@subtelforum.com.

Find us on Facebook

submarine telecoms
FORUM

January:
Global Outlook

March:
Finance & Legal

May:
Subsea Capacity

July:
Regional Systems

September:
Offshore Energy

November:
System Upgrades

Conferences

ICPC Plenary Meeting
12-14 April 2016
Hamburg, Germany
Website

SubOptic 2016
18-21 April 2016
Dubai, UAE
Website

PTC 2017
15-18 January 2017
Honolulu, Hawaii USA
Website

 SUBSCRIBE TO OUR FEED

 JOIN OUR MAILING LIST

Voice of the Industry

CODA
BY KEVIN G. SUMMERS

It's springtime on the farm. The grass is green, the birds are singing, the cows are breaking down their fences, and we're sewing seeds in the garden. It's a wonderful time of year... the weather is so pleasent that it has the opposite effect of winter's seasonal depression. I like to think of it as seasonal mania. I'm running from one project to the next, living on coffee and butter, and here we are again with another issue of SubTel Forum.

I know I've said it before, but my favorite thing about modern technology is that it makes it possible for me to work from home or for Wayne to work from the beach in North Carolina, or for Kristian to work from a hotel room in Dubai. And if a problem comes up, we can all conference together and resolve it. That's remarkable! Think about how much time we as a people used to waste in meetings.

Of course, in order for someone to be able to work anwyere, there has to be an internet connection. I have a little place in New Hampshire that has no internet. The only options in the whole town are satellite and dial-up. When I visit my house in the Granite State, I don't even bother to bring my computer because there's no point. How many locales are left in the world without reliable, high-speed internet? Quite a few, and in most cases, the people that live in those remote areas are often the ones that would benefit the most from a nice fiber optic connection. I do my fair share

of complaining about the 3G connection at my farm, but it's light years ahead of having to drive 35 minutes to a Dunkin Donuts in order to connect to their satellite dish, which is what I have to do in New Hampshire.

What's the answer? It costs a lot of money to install fiber optic systems, and these remote areas are usually sparcely populated, so there's little hope of a return on investment. Let me know what you think? The maple is running and I'm headed to New Hampshire.

Kevin G. Summers is the Editor of Submarine Telecoms Forum and has been supporting the submarine fibre optic cable industry in various roles since 2007. Outside of the office, he is an author of fiction whose works include ISOLATION WARD 4, LEGENDARIUM and THE MAN WHO SHOT JOHN WILKES BOOTH.

 +1.703.468.0554

 editor@subtelforum.com

Voice
of the
Industry